Theatre and Consciousness

Artists and Issues

in the Theatre

August W. Staub
General Editor

Vol. 14

PETER LANG
New York • Washington, D.C./Baltimore • Bern
Frankfurt am Main • Berlin • Brussels • Vienna • Oxford

UNIVERSITY COLLEGE
WINCHESTER

Martial Rose Library
Tel: 01962 827306

2 6 JAN 2009

3 0 MAY 2005

- 6 FEB 2007

- 5 OCT 2009

2 7 NOV 2007

- 4 MAY 2010

- 6 OCT 2008

- 8 MAY 2012

- 3 DEC 2008

To be returned on or before the day marked above, subject to recall.

Gordon Scott Armstrong

Theatre and Consciousness

The Nature of Bio-Evolutionary Complexity in the Arts

PETER LANG
New York • Washington, D.C./Baltimore • Bern
Frankfurt am Main • Berlin • Brussels • Vienna • Oxford

Library of Congress Cataloging-in-Publication Data

Armstrong, Gordon Scott.
Theatre and consciousness: the nature of bio-evolutionary
complexity in the arts / Gordon Scott Armstrong.
p. cm. — (Artists and issues in the theatre; vol.14)
Includes bibliographical references and index.
1. Aesthetics—Psychological aspects. 2. Theater—Psychological aspects.
3. Genetic psychology. 4. Brain—Evolution. I. Title.
BH301.P78 A75   128—dc21   2002021409
ISBN 0-8204-5773-6
ISSN 1051-9718

Die Deutsche Bibliothek-CIP-Einheitsaufnahme

Armstrong, Gordon Scott:
Theatre and consciousness: the nature of bio-evolutionary
complexity in the arts / Gordon Scott Armstrong.
–New York; Washington, D.C./Baltimore; Bern;
Frankfurt am Main; Berlin; Brussels; Vienna; Oxford: Lang.
(Artists and issues in the theatre; Vol.14)
ISBN 0-8204-5773-6

The paper in this book meets the guidelines for permanence and durability
of the Committee on Production Guidelines for Book Longevity
of the Council of Library Resources.

© 2003 Peter Lang Publishing, Inc., New York
275 Seventh Avenue, 28th Floor, New York, NY 10001
www.peterlangusa.com

Printed in Germany

For Jan and for Catherine:
"… A little lower than Angels…."

Contents

Preface

In past years, when I was a much younger man intent on developing my career by visits to Samuel Beckett in Paris (which remain among my most gratifying and fulsome experiences), I stayed at the Hotel Esmeraldo in rue St. Julien le Pauvre, on the West Bank overlooking the Seine and the Cathedral Notre-Dame.

Across the tiny street is a little park, a green swarth facing the Cathedral. To the back of the park lie benches and a sandlot where people congregate; a sanctuary facing the Quai de Montebello; a vista of stone wall and river and nine hundred years of French history, just a short stroll from Shakespeare and Company. Presumably even today ex-patriot writers sit cramped in an upper stairwell, putting down their soon-to-be immortal words.

At night, from my room on the third level, I look out to see the Parvis and Portail de Jugement of Notre-Dame, circa 1163, lit by searchlights. The morning brings a one-legged pigeon that sits patiently as I crank open my window and carefully lay out a crusty stale baguette. Next morning s/he's back. Thereafter I never fail to bring home with me my shared pigeon baguette; so much for nostalgia.

Early on the second evening of my first visit, I descend the narrow circular staircase (I don't know how I managed to get my bags up all the way!) when—approaching the concierge's desk around the last turn—I hear a radio broadcast, at full volume. To my ear, not accustomed to listening to French, it wasn't as much unfamiliar speech as it was singing—nuanced rhythms—that echoed all the way up the stone staircase. Not the words as much as the inflected voices and the rhythms, the actors' speech was almost hypnotic.

Of course we have all experienced this rush of feelings, in any language, as we watch and listen to the delivery by great actors of our time. But what stood out in this Comedie Francaise production were the song qualities of speech—a "hymn to theatrical consciousness." My Paris experience presaged a profound truth about language that a rather remarkable Oxford scholar, Jean Aitchison, recently proclaimed: human speech is more like birdsong than any of the pongids.[1]

In range of expression, in precision and in speed of comprehension, not even our hominid cousins can approach the marvel that is speech. When we talk about consciousness, of course, our feathered friends do not track step-by-step. Con-

sciousness involves a whole different set of parameters. But the functions of the human brain to develop an evolutionary scheme we call mind and consciousness and speech suddenly moves back in time, from 7.9 million years to at least 65 million years before the present. (Human speech is no more than 200,000 years old.) In itself, that new dimension is an astonishing leap.

The basis of consciousness derives from the generative power of a need in human beings to communicate the nature of their sense of the world. That sense has taken a long time to emerge. The nature of theatrical consciousness, which lies at the heart of consciousness, is a product of a reflective and discontinuous cortex. Having read this, you can close this book and move on. Or you can read on about the most fascinating and diversely patterned evolutionary track in the history of earth. This story is nine and a half billion years old.

Acknowledgments

In writing a text that has no predecessor in the field of theatre studies, one is constantly aware of the proximity of the irrational cliff, over which greater and even more speculative men and women have fallen. That has not reduced the ardor and esteem I hold for Gus Staub, Chair Emeritus of the University of Georgia Drama Department, who recommended this text for publication, nor for the persistence of the editors of Peter Lang, for venturing to bring these words to print. Most of all, on the human scale, I wish to give thanks to my evolutionary family: to Lucy (3.9 million years before the present), for her 195 thousand generations of successful begattings; to the earth (4.55 billion years before the present), for her 227.5 million begattings; to our solar system (9 and one-half billion years before the present), for its 4 billion, 750 million transformations, and most of all, to the nth generations of supernovas, whose traces we can still observe in the refraction of our sun's rays, back to the Big Bang (14 billion years ago), for 7 billion transformations.

All of the forgoing would be impossible without acknowledging Murray Gell-Mann, author of *The Quark and the Jaguar*, and a co-founder of The Santa Fe Institute for Complexity Studies. To date there has only been one successful universal theory of the universe, that of Einstein's Theory of General and Special Relativity concerning the functions of gravity on galactic bodies and super novas. Murray posed the existence of quarks as the penultimate particle physics. He then reversed direction and posed the importance of relations of objects as the key function— much as Foucault, in the visual and moving arts, suggested that the relations of objects to their surroundings was as important to theatre theorists as was strata of rock formations to archeologists and geologists.

At some future moment, a true unified theory will embrace the macro and the micro universe, whether it comes as a mathematically-based formulae of a computer-patterned, 26 dimensional string theory, or some particle super quark physics not yet formulated. Paul Davies at the Santa Fe Institute says it better. "There's the reductionist path, in which you try to break things down into their most elementary constituents—quarks, or maybe something deeper, like superstrings. The other path is the path of synthesis, the path of looking at the complex organizational arrangement of things and recognizing that there's a whole science of complexity, with laws and principles emerging at successive levels."

I suspect that one universal field theory will be a patterned rhythm, not a singularity but a bipolar universal yawn, a reverberation, or successive yawns, or even a binary guffaw that may never end.

To all of those who have contributed insights, known and unknown to this writer, acknowledged and unacknowledged in the text, but whose ideas have found credence in the words of those whom I cite, my eternal thanks. This is very much a work in progress.

I would also like to thank in particular Regis de Silva, MD, at Harvard Medical School, for encouragement over the years, for early conversations on a tri-partite model of the brain, and Scott Kelso at Florida Atlantic University in Boca Raton, for his research into patterned responses of brain activity, for my first anatomical lesson in handling a cadaver human brain, and for reflections of a professional child actor in Dublin, Ireland.

Finally, I would be remiss if I did not pay tribute to the political skills of Andi Sutherland, who managed to get me an invitation to Santa Fe, to the brilliant Jim Crutchfield, who took me to the brink of computational mechanics of cellular processes but couldn't convince me that a solution was at hand, and to George Cowan, a resident founder-genius of the Complexity Studies Institute, who responded to my inane inquiries with infinite patience and courtesy.

To colleagues who have encouraged this study, foremost of them Gus Staub at the University of Georgia, Claude Schumacher in Mazenay, France, Nicole Vigouroux-Frey at the University of Rennes, France, to members of the Executive Committee of the International Federation for Theatre Research, who have invited me to speak at Theatre Congresses in Amsterdam and in Canterbury, England, on this topic, and to friends who have listened to impassioned tirades over the years in distant places at obscene hours, my deep thanks.

Part One

❧◉❧

EVOLUTION
OF THEATRICAL
CONSCIOUSNESS

Men vent great passions by breaking into song, as we observe in the most grief-stricken and the most joyful.

—Giambattista Vico, *The New Science*, LIX

Chapter One

❦

The Discontinuous Process

BEGINNINGS

I believe that consciousness is not a thing or location, but a rhythm that begins deep in the old brain, perhaps 200 million years ago. Neural maps that link the *pre-frontal lobes* of the *cerebral cortex* of man to the *thalamus*, whether they begin with the old brain and connect to the forebrain or the reverse, are the most likely sources of the wonder that is self-awareness.

Through the processes of evolution in the last 200,000 years, this potential has emerged as a discontinuous series of processes that we call consciousness. Externally, a short step from self-awareness is the display of self-awareness we call performance. Internally, in theatre the key element is not language or even vision, but movement that introduces change in the brain.

Hence rhythms of performance bit, or figure/ground or light wave/color constraints are crucial for perception. The "absence" so noticed by observers of theatre performance is easily explained as the necessary pause that enables each spectator to reflect and selfishly (in a good sense) augment brain perceptions. More about this topic later.

In a recent review of Semir Zeki, *Inner Vision: An Exploration of Art and the Brain,* and of Donald Hoffman, *Visual Intelligence: How We Create What We See,* Israel Rosenfield noted the Freudian world view of the "eye as camera" has been displaced in the past thirty years by the concept of sight as a selective series of learned disparate processes in the brain. There are now specialized functions for the analysis of different properties, such as color, shape, and movement. The visual world, colors of the rainbow, illusion of motion on the movie screen, three-dimensional space, are creations of the brain.[1]

This suggests that our visual world is not determined by light but by wavelengths, which are "creations" that depend upon complex interactions within the visual cortex. More specifically, the physical qualities of color are the brain's solution to making sense of the constant flux of visual sensations that are registered on the retina.[2] There is a world of difference between fixed images in the brain and rhythms or patterns in the brain, and that is what this text is all about. This kind of analysis opens the door to issues in complexity studies analysis.

"What is Complexity Theory," you might well ask, and "why should we concern ourselves with this kind of research analysis?" The simplest answer is given by Paul Davies, in John Brockman's *The Third Culture: Beyond the Scientific Revolution*. Describing the work of Murray Gell-Mann, one of the foremost particle physicists of the present century, Davies noted the minimalist- and the universal macro-world approaches:

> There are two ways of studying the world. There's the reductionist path, in which you try to break things down into their most elementary constituents—quarks, or maybe something deeper, like superstrings. The other path is the path of synthesis, the path of looking at the complex organizational arrangement of things and recognizing that there's a whole science of complexity, with laws and principles emerging at successive levels.[3]

This text looks at the patterns, the rhythms and the vibrations inherent in the evolution of species in the last one hundred and ninety million years, and the complex organizations that have determined how and why we got to where we are now, rather than the constituent parts—or quarks—at rhe bottom of the nuclear pile.

In self-regulated systems, relations between data are more important than the data itself. Furthermore, if this is true for a generative visual cortex, this is also true of the cerebral cortex, and of brain functions as a whole. Rhythm, movement, evaluation, reflection, are the rhythms of consciousness, whether narrowed to the retina, to general awareness as a whole, or—for our purposes here—to the actions of a stage! The nature of consciousness is process and movement, from reflection in the age of the dinosaurs to the last great ice age that ended only twelve thousand years ago, and thence to modern man.

Research in the evolution of modern man has reached a milepost in the twenty-first century with the employment of computers to analyze complex adaptive systems like the brain and mind of man. Research centers like the Santa Fe Institute promise extraordinary insights into the emergence of modern human behavior during the Upper Paleolithic Period, 45,000 to 35,000 years ago. During this period, archaic hominids disappeared, modern man claimed dominance, technology and civil behavior advanced.

Researchers like Jeffrey Brantingham at SFI are now asking questions: "What were the underlying biological, behavioral, and environmental factors structuring and/or driving the emergence of Upper Paleolithic technological adaptations and modern human behavior? And what organizing principles, both inherent in these complex systems and emergent through their interactions, allowed modern humans to colonize extreme environments such as the High Arctic, the arid core of the Mongolian Gobi and the Tibet Plateau?"[4] At some near date, research will encompass the full span of reptilian mammals, circa 195 million years before the present.

LANGUAGE AND BIRDSONG

While Jean Aitchison's speech text does not give many details of the origin of the species of *Homo Sapiens*, (now circa 6.9 million years before the present,) she does present a theory of why there was a differentiation several millions years after this period. Following the seismic birth of the Great Rift Valley in Africa, Aitchison believed that tribal species were physically separated. To the west went the pongids into the wet tropical climates; to the east went the precursors of modern man, onto the relatively dry and treeless savannas of East Africa.

We now know the linear progression of Homo sapiens' forbears to be quite wrong. Not a tree but a bush; not a single species but a great many variations of species; not one locale but a great many locales in east, west and central Africa, were home to the ape that got lucky.

Life in Africa vastly changed in the period 6–8 million years before the present; climatic changes that fostered great tropical forests became vast grasslands. As a consequence, tree-dwellers declined in numbers, and those who could adapt by becoming two-legged foragers flourished, in turn died out and were replaced by yet more adaptive species. It is noted that 10 million years before the present, apes proliferated. By 7–8 million years before the present, only a few species of apes were found. Within a million years later, human and chimpanzee lineages divided, and the discontinuous brain of modern man, spread across a limited number of hominid species, was caught up in a great evolutionary roll of the dice game. It would seem that creatures with discontinuous brains like ours thrive in adversity, while well-adapted species fail.

Evidence of this evolutionary crap shoot has been uncovered in several geological sites in 2002—an African skull from Chad, central Africa, circa 7 million years before the present, and a Georgian skull, circa 1.75 million years old. In commentary on these discoveries in the *New York Times* of August 6, 2002, Dr. Barnard Wood, paleontologist at George Washington University, notes that "this really exposes how little we know of human evolution and the origin of our own genus Homo."[5] More startling even than the Toumai skull from the Djurab Desert in Northern Chad is the 1.75 MY old Georgian skull of only 600 cubic centimeters (versus modern man's skull of approximately 1400 cubic centimeters.)

The assumption has been for generations that the first out-of-Africa migrants were Homo erectus, a large-brained creature who wandered out into the world with larger cranial capacities, longer legs and more developed tools. Now comes a creature from Georgia, in a small medieval town named Dmanisi. that resembles Homo Habilis, who preceded Homo erectus. It would appear that small-brained adaptive creatures followed other species out of Africa at a very early geological age, and that brains have little or nothing to do with species radiation. The luck of the draw seems more evident than ever, just another example of complexity theory working through evolutionary generation across millions of genetic simulations

(Even as I write, the date of an upright precursor of modern man has receded to 6–7 million years before the present. As we approach the absolute earliest dates, the chances of finding a representative skeletal remains becomes even more problematic. There is every likelihood that the first four-legged creature in this evolutionary line stood up, however briefly, seven million years ago or more, and so startled his or her peers that s/he was given a choice of mate or browsing turf, and so endowed that descendents' line!)[6]

In that prescient moment of "climatic crisis," evolutionary advantage played out a role that seems, in retrospect, almost predestined. With survival in doubt, the discontinuous cortical nervous system of pre-modern man adapted, evolving into a new branch of the species, and flourished. *Homo Erectus* became *Homo Habilis*; hands adept at swinging through trees became hands and arms adept at using tools. It's not so much that "we" were destined to become who we are, but that a creature somewhat like ourselves, taking advantage of climatic and environmental crises, and even galactic forces, uses a reflective brain to adapt—and thereby to survive and to fight another day.

Why not have two hearts, three sets of eyes, wheels instead of legs, or snake-like coils with six pairs of arms? As it is, *Homo Erectus* got there first, thrived in adversity, and did not give up the store.

But what of Aitchison's bird-song and why do we begin there, with birds, the modern descendants of the age of dinosaurs, at least sixty-five million years before the present? The answer is, in part, that birdsong is a universal trait in humans, with instantaneous translation into meaning through adaptation of a gene that is no more than 200,000 years old. It may be that a genetic mutation, an accidental fluke of nature, prepared early modern man for his role in our millennium! Ask the Comedie Francais. Ask historians of the origins of "melodrama." Ask anthropologists to show you an endocast of a human brain! There are no fossils of the brain, only of the brain case. Similarly, 19th-century melodrama bears little similarity to its seventeenth-century namesake. But they sang a song that was not even possible 300 centuries ago.

There is also probably good reason not to speculate why a tiny forest creature, surviving under foot of the last of the dinosaurs by dint of a reflective and discontinuous nervous system, did survive and evolve into man. The evolution is not entirely clear, especially in the period 65 MYbp[7] to perhaps 9 MYbp. One might almost say that the chance impact of a galactic body into earth was the handiwork of an almighty figure. Alternatively, in the course of over seven and a half billion years, evolutionary chance made possible the emergence, through trial and evolution, of a creature that could understand the possibilities of a knowable universe.

The argument is not so much that modern man emerged as the evolutionary victor, but that a creature with a remarkably theatrical brain did, capable of perceiving what we now know as theatrical acts. With a discontinuous cortex, this outcome was a given. What is remarkable is the form this creature finally took, given the choices that emerged after such a lengthy gestation. One might finally say that it took seven and a half billion years of experimentation to create one possibility of a dominating discontinuous cortex, composed of equal parts of guile and godhead.

Why did it take so long? Isn't theatre the basis of this species that can dispassionately comprehend this version of a creation, existing only at the speed of light in a time/space continuum? And why this species and not another, given the options of adaptive success and failure across that continuum? What would happen if we move outside of time and space, into a "non-state" that has no space? If there is dark matter and unseen dark energy that seems to lie outside our physics, there must be another physics, a "mother of all physics" holding the universe together. And if all the galaxies don't have enough mass to generate the gravity to hold these galaxies together, does the dark matter supply the gravitational glue? According to current physicist's calculations, 85% of the universe is missing, or unseen, or collapsing into black holes at the center of our own Milky Way.

If space bends, what does it bend towards or from, and why should there be a bend if there is nothing without, in which to contain that bend? If there is a galactic force there must be an impetus and why an impetus out of which a force has the speed of light? Why does energy exceed the speed of light at the edges of the universe. And why is the Big Bang a singular events, and the universe deemed to be a singular event of expanding into infinity? At the very least there should be a binary rule, and a billion Big Bangs, and a duodecillion universes expanding infinitely beyond non-time and space.

Why is 85 percent of the universe—comprised of dark matter and dark energy—unseen, while its gravitational effects are seen? Is the subatomic world also a geometric world? Is the speed of light slowed by dark energy to 186,000 miles per second everywhere in the universe? Why does space-time begin after the Planck era, and contain 10 dimensions during, and no dimensions before? The list goes on. We can reflect and imagine, we can visually create—as our ancestor did in the age of original birdsong; hence "theatre" is a very old phenomena that likely predates our known species' origin. I would postulate that discontinuous reflection is the origin of our evolutionary success. Evolutionary success is a product of the origins of theatre.

The great problem for researchers in consciousness and in my particular case, theatrical consciousness, is the lack of evidentiary proof that such and such occurred at a particular moment in time. Beyond slight indentations in a skull, the brain leaves no trace of its activities. We must look elsewhere, historical records and the like, to find proof of intelligence. But Professor Aitchison's research has uncovered a Rosetta Stone of information. Every time we open our mouths, we give evidence of our origins. If we have the rare insight to track the evolutionary changes across the world, for the last fifty thousand years, the migratory patterns of evolutionary *Homo Sapiens* appear. Hence, coupled to linguistic clues for the last fifty thousand years, we can track the fallen larynx of Modern Man back some two hundred thousand years and the adaptability mechanism of precursors back sixty-five million years.

Perhaps the most telling proof of the existence of a language gene is work done in 2002 at the Max Planck Institute for Evolutionary Anthropology in Leipzig, Germany. Earlier research showed that a mutations in a gene, called FOXP2, caused a wide range of speech and language disabilities.[8] The research group traced this language gene back through several primates—chimpanzee, gorilla, orangutan, rhesus macaque, and the common mouse. In the common ancestor of humans and mice, dating back 70 million years, Savant Pabo's group noted only three changes in the protein's amino acid sequence—with two of the changes in Homo sapiens coming since man split from the pongids. In dating the last amino acid substitutions for modern man, the research notes are fascinating:

> Although the date cannot be pinpointed, the team concluded that the fixation was 95% likely to have occurred no more than 120,000 years ago, and virtually certain to have occurred no earlier than 200,000 years ago.[9]

Clearly, the dates could not be more precise or complementary to the thesis now being developed. The only question remaining is "the chicken or the egg" hypothesis: given that the speech gene conferred an evolutionary advantage, does this mean that the spread of anatomically modern humans was driven by the evolution of language, or was this gene (which assists in making mouth and facial movements essential to speech), selected because it improved vocal communication once language had already evolved? It might be noted that Neandertals are believed to have had poor communication skills, and may have faced extreme social ostracism as a consequence. That argument would favor the latter thesis, of modern man moving out from Africa into the world with a new language gene and therefore, an infinite evolutionary advantage.

The final ingredient for success is outlined in Aitchison's work, which validates speech as the end product of a thinking, reacting brain. In each succeeding generation, genetic codes mandate adaptive complexity responses during the maturation processes, including the discontinuous reflective processes of speech and thought. This latter capacity is an extraordinarily interesting mental mechanism that guarantees evolutionary success if the physical species can survive. In summary, as a species we flourish because our genetic codes blueprint an evolutionary path that is theatrical to the core.

Combined, these parameters of modern consciousness have visible origins. But to find the real origins of theatrical consciousness, and trace those origins through the emergence of bird-song in man, we must go back to the dinosaurs. Somewhere in that period "beyond time," as Disney World would proclaim it, a creature with a discontinuous cortex, a "not hard-wired" brain, began its reflective ascent up the dominance ladder of earth's species. What did it reflect? Movement! Mimesis: Aristotle's definition of theatre as "imitation of an action," or perhaps, "imitation of movement in its symbolic mode" can be hazarded. Connecting up the picture dots to date, let Aitchison be a guide from the valley of prehistory.

From the first instance, Aitchison notes the movement factors in language. She even cites what she and others have called "generativity" as the basis of what we call mind. The discontinuous organization of the brain allows the mind infinite opportunities to create infinite numbers of forms and sentences.[10] The element most interesting is the uniformity of emergent possibilities across the globe. This complexity process, linking birdsong and human speech, is the quantum leap of humans over other species: the grammar is precise, uniform and without confusion *without exception.*[11] Most important, language communication is also instantaneous.

Aitchison reminds us that we are all descended from a fairly small stock of African cousins, circa 4 myBP. Prior to that we are hominids, a division of primates who had diverged from apes some 6 myBP. We brought with us the birdsong heritage, and we have remained the only species that talks in birdsong. The others—panids (chimpanzees), and pongids (gorillas) remain our closest relatives—have language prospects that are not generative.[12] A graph for primitive mind generation, speech generation, and language diffusion is, as follows:

350 myBP	*Crossopterygian;* fish-like predecessor of reptilian mammals
195 myBP	first mammal, *Hadrocodium wui*
65 myBP	species forebear; reflective discontinuous cerebral cortex. Dinosaurs disappear
7–8 myBP	pongid and hominid species split
6.9 myBP	first upright man-ape forebear, *Sahelanthropus tchadensis,* in Chad, "Toumai."
6 myBP	*Orrin tugenensis,* in Kenya
4.7 myBP	man-ape forebear remains found
4.5–4 myBP	*Australopithecus ramidus,* in Ethiopia
4 myBP	*Australopithecus afarensis,* "Lucy"
3 myBP	*Homo* species
2 myBP	*Homo habilis:* "handy man" tool user
1.75 myBP	Dmansis *Homo habilis* "out-of-Africa" species
1.5 myBP	*Homo erectus:* "upright man," descendents of Rift Valley split, move into Eastern Africa, Georgia
300,000 yBP	Archaic *Homo sapiens*
175,000 yBP	*Homo sapiens:* modern man with fallen larynx and capacity for human speech
120,000 yBP	Dispersal of *Homo* populations out of Africa
80,000 yBP	*Neandertal* burial sites with symbolic rites of passage
50,000 yBP	Middle East, Asia colonized

45,000 yBP	Europe colonized
38,000 yBP	cave art, France and Spain, Aborigines arrive in Australia
30,000 yBP	The Americas colonized across Bering land bridge
12,000 yBP	tribal civilizations
5,000 yBP	formal language
3.200 yBP	organized theatrical events, Egypt, Crete (?)
2,300 yBP	earliest written surviving theatre epic, *Gilgamesh*, from Syria

The specifics are more interesting to anthropologists. The only thing I would note here is the scarcity of monuments to cognitive activities in the early record, the great rush of significant communication changes in the last 35–50,000 years, and the communicative gene for humans that arose through punctuated equilibrium perhaps, 200,000 to 120,000 years ago. In this regard, it cannot be stressed enough that luck has played a major role in our emergence as a dominant species. Eight million years ago, the forests of Africa were filled with myriad species of apes. Then life changed, forests shrank, species disappeared, animals emerged that could deal with change, and *Australopithecus afarensis*, "Lucy," flourished by chance—the ape that got lucky.

Language and generative mind processes, which had lain dormant for millions of years suddenly, exploded. Now the most routine communications, as natural as walking out to one's car, were literally impossible epic adventures only a few short thousands of years ago!

There is a lesson of history in all of this. In fact, there may be as many as five galactic history lessons. Mass extinctions, caused by asteroid impacts with the earth, have occurred 420 and 350 million years ago, and as recently a 250 million years ago, on the boundaries of the Permian and Triassic periods. A fourth extinction occurred 199.6 million years ago that led to the ascension of the dinosaurs and the mass extinction of mammal-like reptiles.[13] The repetitive catastrophic events of a meteoric strike of earth on the Yucatan Peninsula, 65 myBP; the tectonic events that created the Great Rift Valley several million years before the present; and the devastation in the last great ice age, 125,000 years before the present, down to about 16,000 IBP; have forced *Homo sapiens* to adapt or to perish.

Humans have adapted and flourished, almost as if they were following a DNA genetic code: preplanned, and read by their discontinuous cerebral cortex. In the meantime, the rationale for survival seems to be not merely adaptability but excessive good luck. To survive on this planet is to win the celestial lottery every 100 million years. There are more subtle influences, of course. Aitchison notes that *australopithecines* remained vegetarians, while the *Homo* species added meat to their diet and thereupon developed a more nourished and larger brain. She also suggests that language began around 250,000 years before the present, somewhat earlier than the esophageal changes, and developed rapidly until around 50,000 years before the present, when a period of relative stability set in, and conditions for civilizations began to arise.[14]

This particular view would corroborate a "punctuated equilibrium" view of Jay Gould whereby modern man, isolated by the ice age, developed in a small family tribe an aberrant left hemisphere brain function. This function, known as the *secondary association area* of the left *angular gyrus*, proved so efficient in generating sound images that it became a key strategy for survival throughout the species. Hence in an evolutionary trice, the basis of human language was born. Capacity for developing a grammar of sounds and images during the formative years became standard, and reinforced with each generation, until a reasonable capacity was reached about fifty thousand years ago.

There is some prospect of closing the operculum of the left hemisphere (as in the case of Albert Einstein's brain), but that may take another fifty thousands years, when perhaps an abstract digital right hemisphere brain may operate across the *corpus callosum*.

At the present time, no one knows the future evolutions of approximately 37 mitochondrial DNA, located in the mitochondria structures, in addition to the regular complement of 100,000 genes, located on DNA molecules within the nuclei of human beings' multimillion cells. These "DNA instruction packets" are plainly much too complicated and divisive for this brief commentary on the evolution of language, movement, and consciousness.

Given all this ancient prehistory, what is the evidence that language can be shown to play a role in the emergence of consciousness, and in particular, theatre consciousness? There are no artifacts; no fossil remains can be found, and the evidence of "civilizations and its remnants" is too generalized to be useful. This is where Aitchison shines. She has shown it is possible to track the parts of language evolution across the globe in a particularly meaningful way. Her arguments begin with a summary of the three elements of any language: "phonology (sound structure), syntax (word association), and semantics (meaning)." She cites authorities to bolster her argument—the absolutely key argument—that, whereas sound structure and semantics do not require consciousness, indeed, can be the basis of any voice mail system or other machine duplication process, syntax (word organization), is a very thought-full process.[15]

Syntax is a conscious evolutionary factor, and by looking at word order in languages, we can track origins and spread of language, and therefore of consciousness, in many parts of the world. In concept, this analysis is elegant and wonderfully straightforward. Roman Jacobson, an eminent grammarian, suggests a sense of maintaining and breaking equilibrium are indispensable properties of any formal language. This corroborating description is a close approximation of the definition of a complex system. If we add the notion of chaos theory to this description, "sensitive dependence to initial conditions," the evolutionary track back and forth, while never repeatable, can be nicely recorded. For the first time we have hard evidence of a complex system called human language, poised always on the brink of instability.

How does theatre fit into this puzzle? The evidence suggests that language developed rapidly from about one hundred thousand years, to fifty thousand years ago. Emerging from a relatively isolated geographical area (a development called speciation), language-bearing humans had enormous communication advantages in social organization, food gathering and survival techniques, and in making battle plans. Hominid groups quickly overpowered non-language groups and the offspring of defeated groups were incorporated into a "language culture." It is probably evolutionarily significant that only pre-puberty *Homo Sapiens* have this capacity. This suggests that adults of a defeated tribe would have been slaughtered and their young would survive by learning to speak syntactically.

This hypothesis is only possible if it can be assumed that a structure or series of structures in the mouth, larynx and brain evolved by a series of chance, or by an emerging capacity for consciousness that generated a coordinated response toward communication. In this proposed scenario, theatre lies at the origins of modern man. The reflective brain demands a means of communicating this reflectivity to members of an immediate tribe or family.

In summary, theatre is not a product of consciousness; rather, consciousness (and language, which followed many thousands of years later) is a product of a need to communicate to others the processes of a reflective, discontinuous brain. Why this all took some sixty-five million years to emerge, and countless species through which to evolve, is probably the onion skin of a process that took seven and a half billion years to finally "get it right." The available explanations of the emergence of the universe, the emergence of consciousness, the development of human speech, have awaited an emergent science of complexity. When we develop a Standard Model unified theory or find that string theory is yet another aspect of complexity theory, where symmetry without causation, or a geometry of vibrations is the without of the quantum vacuum.

PRACTICAL LIMITS

As a result of late twentieth-century scientific breakthroughs in pure theoretical physics and mathematics, in practical research in neuro-anatomy, and in climatology, the concept of chaos as a fundamental law of the universe became widespread. But this could not explain how a system could maintain itself and adapt beyond the initial inertia.

Within a decade, chaos theory (sensitive dependence on initial conditions) has been displaced by complexity studies (functions of a self-regulating system). Complexity theory also suggests that internal modeling and adaptations to natural forces are always probabilistic. Given certain parameters, the results will vary within a predictable range. This kind of reasoning has the immense advantage of looking at natural forces in the natural world. Whereas in classic physics, the mathematics assumed that the presumed environment was a vacuum, in complexity studies, the mathematics has become computer-modeling simulations where the real environment is part of the sensitive dependence on real life conditions. Uncertainty is built into the system.

In an issue of *Complexity*, Murray Gell-Mann explained some of the problems in terms of algorithmic information content (AIC) or the "shortest program that will cause a standard universal computer to print out the string of bits (of information) and then halt."[16] He goes on to explain that most of the effective complexity of the universe lies in the AIC of a description of [accidental conditions of probabilistic events.] The evolution of our species, for example is a condition of chaotics, and complexity, all at once:

> Any entity in the world around us, such as an individual human being, owes its existence not only to the simple fundamental law of physics and the boundary conditions on the early universe but also, to the outcomes of an inconceivably long sequence of probabilistic events, each of which could have turned out differently.[17]

What is striking in the nature of man, and in the nature of the mind of man, of which theatre is a feature, is the tendency of the nature of complexity to expand. "even though any given entity may either increase or decrease its complexity during a given time period."[18] According to Gell-Mann, the life of objective forms will cease, "self organization becomes rare, and the envelope of complexity begins to shrink."

Another milestone in science occurred in 1992, when James Crutchfield designated "Computational Mechanics of Cellular Processes" to address unsolvable problems in conventional turbulent flow systems, or emergent structure in state space, for example, by means of computer simulations. In neuroscience, investigators could probe performance data in the human brain by using functional magnetic resonance imaging techniques. Injected agents in the blood stream were passed through the microvasculature, where magnetic field distortions were produced that were linearly proportional to the concentration of contrast agents, which in turn, were a function of blood volume. This solution may not be the Holy Grail of neuro-research. George Cowan has pointed out that in 1948, Warren Weaver (inventor of the term *complexity*), "distinguished between 'disordered' complexity, characteristic of large assemblies of identical objects with behavior that is described in statistical mechanics, and 'ordered' complexity, observed in systems with nonidentical, interacting objects that behave in unpredictable or only qualitatively predictable ways"[19]

The prescient "Father of Complexity Studies" also predicted that "research on systems with 'ordered' complexity would become a major concern of science in coming decades."[20] Crutchfield used "disordered" complexity methodology to solve an "ordered" complexity problem. Additionally, there may be similarities in neuronal signals across synapses, but we are dealing with sound/image receptors that lead to a whole secondary range of neural mapping. The qualitative axes of the latter are of "ordered" complexity. This is not checkers but three-dimensional chess. Theoretically, "deep structures" of thought processes could be measured in the *primary auditory cortex*, the *cerebellum*, the *visual cortex* in humans, to test mental imagery, speech perception, visual recall, attention modulation, spatial memory, working memory, and word generation, among other topics.

What could not be measured were the rhythms of consciousness, nor the freedom in the interstices of spoken thought that so absorbed artists like Samuel Beckett from the earliest days of his career, from *Proust* in 1931 for example, as Beckett explored the "top-hatted ministers of Time, Habit and Memory." In the neurobiology of human performance, three (and possibly four) constituent elements are of some importance: transformation, compression, neural chaotics, and species interbreeding.

For theatrical purposes, transformation of images refers initially to the power of the imagination to define new boundaries of communications for the spectators. Visual and aural cues of the actor, and the stage machinery that go into the processes of creating the basis for communication, can stimulate the redistribution of neural maps of the mind. When Woyzeck hears the Freemasons underground, when Othello twice "puts out the light," when cell-bound Segismundo "dreams his dream" of a world that was not a dream, in a stage space that is a dream, we know and engage those revelatory moments from a bank of like-memories that can never match the stage depiction, but that appear to make scripted language comprehensible. Our awareness factor is suddenly raised. But from where and whence did these facilitating memories arise?

In *The Anatomy of Memory*, a *Scientific American* report by Mortimer Mishkin and Tim Appenzeller, some answers to the mechanisms of human memory, crafted from 100 billion or so nerve cells in the brain, are beginning to appear.[21] Developing a thesis from the investigations of sensory data along the visual pathway, Mishkin was able to show that there was increasing complexity of neuronal response as the pathway moved deeper into the brain. In the opinion of these researchers, this kind of global branching at the source seemed to be the general rule for much of the brain's insurance mechanisms. Investigation has now shown that two structures seem to be paramount in laying down memory: the *hippocampus* and the *amygdala*, with the latter organ absolutely indispensable. Modifications of this statement remain to be worked out. If both are damaged there can be global amnesia, whereas there is some retention of memory if one is intact.[22]

The applications of these findings to theatre became even more startling with Mushkin's discovery that even if both organs are destroyed, this total loss did not eliminate old memories, stored at some earlier site in the brain. Parts of the *diencephalon*, at the center of the primitive brain, are organized into two structures known as the *thalamus* and the *hypothalamus*. In a link to the semiotic intuitions of Patrice Pavel, research in brain function has now revealed that memory circuits, feeding back on sensory pathways, are circular: That feedback presumably strengthens and so perhaps stores the neural representations of the sensory event. Synapses in the neural assembly can preserve the connection pattern and transform the perception into a durable memory.[23] The result is that the laying of new memories or new association memories, or the triggering of old memories to new association memories, lies deep within our brain, and is not a function of cerebral processes.

It seems apparent the key to language in man is the association of two non-limbic associations in the *cerebral cortex*. Hence, the complexity of language and gesture communication in the theatre, or in real life, involves the new and the old brain, in profoundly complex, interacting neural pathways. In the case of speech in the neo-mammalian brain, these links have existed for two hundred thousand years, in the case of emotion and memory in the proto-reptilian brain, with no cerebral cortex; for over two hundred million "silent" years.

The remaining question of great importance to theatre performers is why, of all the impressions that are received by the brain, are certain neural representations remembered and all others forgotten or selectively adjudicated as of more or less importance? That is, assuming consciousness and attentiveness for the moment, how is any pattern retained as memory?

The answer would appear, in part, to be based on chemical responses. The *amygdala* and the *hippocampus* have connections to the *basal forebrain*, which has the ability to send acetycholine-containing fibers back to the *limbic* structure and to the *cortex*. This chemical release initiates a series of cellular steps that can modify synapses in sensory tissue, strengthening neural connections and transforming the sensory perception into a physical memory trace. But the *amygdala* has an even more important role to play, a role strikingly parallel to discoveries by the late Norman Geschwind at Harvard that non-limbic *secondary association* areas above the *angular gyrus* of the left *temporal lobe*, are the site of sound-image connections, from whence human language arises.

Investigators have discovered now that the *amygdala* has extensive connections with all the sensory systems in the *cortex*. The *amygdala* also communicates with the *thalamus* along sensory pathways that are related to memory. The *amygdala* is also the necessary link to the *hypothalamus*, which is thought to be the source of emotional responses. In other words, the *amygdala* is the clearing-house of the brain, the quintessential gatekeeper, without whose assistance any "rational" or even "irrational" neurobiological enterprise collapses.

Given these conditions, where the *amygdala* plays a pivotal role in directing the association of neural impulses, it is now clear how sensory experiences acquire their emotional weight. As Mishkin suggested, "the *amygdala* not only enables sensory events to develop emotional associations" but, releasing opiate-containing fibers in response to emotional states generated in the *hypothalamus*, this storekeeper-organ enables emotions to shape perceptions and the storage of memories.[24]

In the case of theatre, by triggering the *hypothalamus* through gestures or through script, the actor draws forth emotionally charged events that make a disproportionate impression. Scenic crisis, climax, resolution, and character charisma are all borne of the *amygdala*, working at one remove on the *limbic system*—the *hypothalamus* in particular—and its reciprocal effects on the *cortex* of the actor and thence, in like manner on the esthetic sensibilities of the spectator.

Does this create the semblance of character and stage situations and context? In Robert Wilson's *Civil WARS*, the imposition of a series of sound cues with visual cues overlays follow parallel courses. One sound cue is placed over a completely unrelated visual cue, to obtain a novel mix at the level of primal distribution in the limbic system. At every "intersection," boundaries of re-entry neural mappings of the brain are challenged.

AUDIENCE TRANSFORMATIONS

How does theatre come about? Given these neural arguments above, it is insightful to listen to the practical, working perspective: two ladies of the current stage—one French, one African-American—propose their own elegant and independent responses:

> Theatre begins when one says of an actor that it's as if he's dying but he isn't really dying, it's as if he's walking but he isn't walking. Because if he walked in the same way in real life and on the stage it would look as if he were ambling.
>
> Ariane Mnouchkine[25]

The second response comes from Anna Deveare Smith, reflecting on the results of a Ford Foundation three-year project with Robert Brustein's American Repertory Theatre and the W.E.B. Dubois Institute at Harvard. On the interactions between artist and audience, she has this to say:

> What I want the audience to experience is the effort that it takes me to move from me to the other. Because in that move…is a suggestion that we are not incarcerated in our own identity politics; in that lives the possibility of us.
>
> Anna Deveare Smith, November, 2000[26]

Mnouchkine looks at the nature of the actor transforming into a character, from a spectator's perspective, while Smith suggests the importance of transformations as a freedom the actor confers on the spectator. Both perspectives are enormously enlightening and absolutely acceptable. But neither spokeswoman deals with the generative power invested in theatre, the need to satisfy an almost palpable desire to confirm and reaffirm community values and our place within that metaphoric space.

In 1985 I traveled to Glasgow to deliver a paper at a conference. Before I reached my destination, I stopped in Paris to have lunch with Samuel Beckett. Since my paper was about the playwright and Jack Yeats, the painter brother of W. B. Yeats, I brought a copy with me to get Beckett's commentary. On the whole he approved. What was particularly interesting to me, however, was a portion of my conclusion that the nature of theatre *lay in the interstices of language.* "Yes, That's exactly right!" came Beckett's response, and he emphasized his words by pointing with his finger to the manuscript.

When we analyze theatre it is not so much that something always remains, as Herbert Blau once noted rather remarkably, but that those remnants are a core of evolutionary change beneath the cerebral cortex. Human intelligence in the species has a proto-reptilian origin that is at least two hundred million years old. The neo-mammalian foundation itself dates back at least 65 million years. While a good portion of this discussion can be objectively described and neuro-anatomically deciphered the process of human cognition and awareness can only be stated in relative, Einsteinian terms.

The core of caring for others, siblings, fathers, mothers, newborn children, and for the death of loved ones, transcends the human species. However, the concern for those passing to another realm of existence is a uniquely human trait. If we go back 195 million years ago to the first mammalian species, or to 65 million years ago, to the end of the dinosaurs, there is no trace of afterlife beliefs. There is no trace in early man, 4.7 million years ago, or even in the first modern man, some 200,000 years ago. Curiously enough, the strongest evidence of burial rites comes from Neandertal groups, a subspecies that did not transition to the modern era, and Aboriginal man in Australia, that did. They leave a legacy of caring that may have been their genetic gift to all of us.

Explorations of caves in Israel in recent years have turned up evidence of Neandertal remains that are highly significant. Amud cave, high in limestone cliffs, rising above a dry streambed that leads to the Sea of Galilee, 14 miles away, is a burial site. (See cover.) 50,000 years ago a 10-month old Neandertal child was buried here in a small niche in the cave's north wall. The jaw of a red deer (or possibly, the meat and bones of the deer's jaw) were placed on the infant's hip in a burial rite. Similar offerings have been found at Qafzeh Cave (deer antlers placed on the chest of a child), Skhul Cave (jaw of a boar entombed with an adult), and at Dederiyeh Cave, north of Aleppo, Syria (piece of flint buried with a 2 year old Neandertal child) during the period between 120,000 and 90,000 years ago.[27]

HISTORY MARKERS

The pursuit of culture and civilization is sacrosanctly grounded in the scientific method. However, the acquired knowledge of the last three hundred years deserts the Book of Nature, on which it is originally founded, for the world of objective analysis. As ever more difficult exceptions arise, new paradigms take their place to explain exceptions. Almost immediately there is a shift in moral vision from an epistemological basis for the arts that is founded on scientific explanations of natural phenomena, to scientific "objectivity" that is largely divorced from the book of nature. Only in the last two decades, following Edward Lorenz's obscure paper on meteorology in 1963,[28] has a new computer-enhanced science of complexity begun to emerge.

At the beginning of the twentieth century, Schrodinger's Quantum Mechanics emerged, with little explanation (or even theoretical justification.) In the last quarter of this century, Chaos Theory also developed, seemingly defying logical explanation, but symptomatic of the over-all malaise of three centuries of objective science. To the surprise of many, who see in the conditions of extreme dependence on initial conditions a paradigm for quantum leaps in biology, or quantum leaps in the origins of the universe, chaos theory provides a model. We can restructure the world of culture and the book of nature.

However, there is a huge problem with chaos theory. Someone or something has to give it a shove to get it started. The generative agency is missing. Teachers may jump-start a student to follow an argument in a classroom, but the essential consciousness mechanism is already in play. In the summer of 1995, an intriguing new aspect of complex adaptive systems appeared. Stuart Kauffman's text on self-organization as a basic principle of nature argues: "What we are only now discovering is that the range of spontaneous order in nature is enormously greater than supposed."[29] Looking at the origins of life on earth, for example, Kauffman speculates that complexity itself triggers self-organization:

> If enough different molecules, for example, pass a certain threshold of complexity, they begin to self-organize into a new entity, in this case a living cell. This replication of creation counters what I have previously described as the vastly improbable nature of our existence as a species, given the options.[30]

For Kauffman, linear dynamical systems derive largely from deterministic, classical mechanics, and are relatively simple. However, nonlinear indeterministic systems derive from statistical mechanics and bring us to the edge of complexity studies.[31]

We can predict after the fact but we cannot measure nor ascertain a particular moment before the fact. Computer simulations of millions of trials are possible indicators of a trajectory or rhythm or other dynamic. Similarly, the evolutionary track that leads to human consciousness cannot be plotted except after the fact. We know about species evolution. The brain of man is an evolutionary fact; but the mind of man is a study in complexity theory that is harder to measure than the path of stars in the Milky Way. The same is true of the nature of human language and theatrical consciousness, and even of reflection on movement.

Chapter Two

❦

Complex Solutions

PIONEERS OF COMPLEXITY AND MIND

Human language can now be traced back to about 12,000 to 15,000 yBP. Philip Ross suggests that the Indo-European language descended from a single speech ancestry last spoken about 7,000 years ago.[1] Located in the area of ancient Mesopotamia—eastern Anatolia (now a part of Turkey) and the southern Caucasus (in Russia's Georgia)—this proto-language developed about 7,000 years ago, and spread eventually to encompass some 200 families of languages.[2]

The preparatory stage encompassed some 90,000 years. Then quite suddenly, and uniquely among the species of earth, *Homo sapiens* developed language and speech as a biological urge of their enhanced left hemisphere sensory apparatus.

It might have begun earlier. The development and maintenance of this chance mutation is strong evidence that there were "brain mechanisms" that allowed for decoding of encoded speech. What probably began as a mutation that increased the ability of some to deal effectively with dangerous ice age climatic changes: reflective capacities in the pre-frontal lobe and sub-language capacities in secondary association areas of the *angular gyrus*, opened the door to discrete, fine-tuned, vocal communication.

Emerging from the ice age, *Homo sapiens* were prepared for the next major step: sophisticated written communication with others of the species. *Homo sapiens* communicated in language patterns not because they felt impelled to speak in a way unlike any other species but because their physiological state was best expressed in a function that we now call human speech. A parallel example is proto-reptilian adaptations to flight in the Age of Dinosaurs. Having developed primitive forms of wings in succeeding generations, over tens of thousands of years, proto-reptilian *chance* survival adaptations through flight became the *fixed* basis for survival of the new species.

With the end of the Age of Dinosaurs (65,000 yBP), and the middle-latitude glacial retreat of the last great Ice Age (10–12,000 yBP), certain advantageous genetic accidents became determinants of any new species' survival. Hard-wired dinosaurs could not adapt. Certain hard-wired proto-reptilian birds, small in size, did adapt. In the case of the precursors of *Homo sapiens*, the greatest determinant was not hard-wired strength or refinement of a specific trait, but the species' very adaptability to change—any change to which a highly intelligent species might, upon reflection, accommodate. The art of the theatre is primordially the art of reflection!

Affirming this thesis, Derek Bickerstein suggested that neural changes in the brain were originally designed as adaptations to the environment that proved immensely profitable to the species.[3] What Bickerstein called "secondary representations" can, in the neurology of the brain, be identified as Norman Geschwind's *secondary associations* of the *inferior parietal lobule*, the site of language creation and reception.

The structures that compose the *left angular gyrus*, including *Wernicke's area*, and the *arcuate fasciculus* that connects the *secondary association areas* to *Brocas' area*, constitute the speech mechanism known to exist only in man. Emerging as an incidental issue of survival through the ice age, this genetic aberration, possibly no more than 100,000 years old, triggered a communicative species.

With the emergence of the modern *supralaryngeal tract*, circa 120,000 yBP, and the enlarged *angular gyrus* in place, sounds could be produced and meanings might be attached to vocalizations of a primitive language. Which came first is impossible to say. Lieberman suggested in his research that the completely modern supralaryngeal vocal tract appeared about 100,000 years ago in the Jebel Qafzeh VI and Skhul V a fossil from Israel.[4] With a "modern" fallen larynx, with syntax—the ordering of words that defines human language—resolved in the *secondary association areas,* and with reflective thinking processes possible, writing, modern speech, dramatic representation, could take place: at once, theatre and civilization as we know it became a rational possibility.[5]

There is extraordinary irony that, without the ice age, modern man would probably still be dealing with fire and flint stone tools as the highest achievements of a branch of *H. sapiens neandertalensis*. The development of modern art-producing man is less than one hundred thousand years old. Equally ironic is the fact that, perhaps in the absence of any need to adapt to a new existence and an opportunity to do so in an isolated environment, modern man might not be here at all. There was no reason for his presence.

As Jay Gould has repeated so often, if we were to replay the tape of the biological history of the earth, we would never repeat the present configuration. Without a saltatory leap in evolution, producing a self-reflective species that could project a consciousness onto the environment and name it as a consequence of innate abilities, with a capacity for speech and writing as a byproduct of environmental adaptation, man would—and could—not have evolved.

We have human language because we have a particular type of brain, of which language is its natural phenomenological pattern. The ability to decode vocalizations has become a natural function of our brain, as natural as the ability and desire of a colt to run. Noam Chomsky amplified this point, acknowledging that we seem to be pre-programmed to learn flawless language as a child and language with an accent as an adult.[6]

Undoubtedly, the evolution of human language followed the evolution of the brain, each reinforcing the other through successive discontinuous selective generations of the species. It is little wonder then that Heidegger would mistakenly claim that "language is the House of Being" and that Being requires that the various features of Being be reflected in the structures and characteristics of language.[7] In *Being and Time,* Heidegger saw the successful use of phenomenology, "to be inextricably bound up with an adequate understanding and insightful use of language."[8] However, despite the claims of phenomenologists, the product is not the process. We don't need language in order to have thought. Norman Geschwind's neurological research at Harvard Medical School effectively demonstrated patterned thought.

Looking at language and the brain of man, Norman Geschwind decided that the function of language and the creation of human concepts could not take place without the presence of an enlarged *left angular gyrus* in the *temporal lobe/inferior parietal lobe* junction. No other species on earth has such an enlarged area in its brain. At this site, "secondary association areas" connect visual stimuli with sound stimuli, and transmit those impulses via the *arcuate fasciculus* to the motor area of the frontal lobe.

Of interest to computer simulations, the human brain works in parallel sequences, with predetermined "patterns on patterns" that fire as a unit—and not singularly. This neural mapping creates verbal thought patterns and communication. To embody these "secondary association areas" with a physical presence called "mind" is the presumption of the species!

THEATRE AS AN ADAPTIVE SYSTEM

The nature of theatre is central to the nature of the species. The adaptive nature of man is based on a discontinuous, reflective psyche that began millions of years ago. The moment that a tiny forest forebear, 65 million years before the present, began to ponder survival, the basis of theatre was born.

Research in 2001 has pushed that date back to 195 million years before the present, in southwest China. A tiny creature who could have curled up on a half-dollar, may be the ancestor of us all. Paul Recer describes this special breed of reptilian mammal—the *Hadrocodium*—discovered by Carnegie Museum of Natural History researcher Zhe Xi Luo:

> ...this tiny animal, like most mammals of the era, probably hid during the day, when dinosaurs prowled. Mammals had better eye sight than did the dinosaurs and could maintain a constant body temperature in the chill of darkness.[9]

Luo further notes:

> A group of animals called "mammalian reptiles" split off some 280 million years ago from
> the animals that were to evolve into dinosaurs. Later, another split led to true mammals.
> The anatomically recognizable mammal came about some 230 million years ago, in the
> late Triassic.[10]

According to Luo, the *Hadrocodium* was the most advanced reptile in that era. We
might now say that the *Hadrocodium* is the first theatrical mammal yet known on
planet earth.

The essential nature of man as a theatrical species is the ability to reflect upon
a discontinuous world. Every performance since that date is a replication of limbic
system responses, moderated by cerebral instructions. Were we to truly analyze
theatre performance, we must begin at that point and work our way forward. Fine-
tuned *fight, flight, sexual encounter* and *hunger* is the entire theatre repertoire; there
are no other options. Our theatre descriptions of character, fictive stage events, historic events, methodologies, are secondary. To reduce the study of theatre to the
"Dilthey dichotomy"[11] has proven to be troublesome in the extreme.

Through the use of complex adaptive systems (CAS) theory which seeks to resolve apparently impossible tasks for linear scientists by suggesting that non linearity is the basis of the universe, no less than the basis of theatrical enterprises
everywhere, we can make a quantum leap in understanding the dynamics of our
comprehension of the fictive space of theatre.

Quite possibly, the self-organization principles of evolutionary biology that led
to modern man were at work all along, creating diversity and high intelligence precisely because we are the most adaptive species on earth. Life arose, argued Kauffman, not from accidents in the bio-evolutionary history of this planet, but from an
expected fulfillment of the natural order. (The verdict on this issue is not in: Harvard's Stephen Jay Gould, for example, would disagree with Kauffman!) It is characteristic of humans that they exist reflectively in the consciousness of their own
existence. It is characteristic of a stage production that spectators partake of the fictive consciousness of the characters. To the degree that this enterprise is successful,
the production will be judged a success or failure.

As G. B. Madison pointed out, modernism insisted that man is a body endowed
with consciousness.[12] Unfortunately, no one defined just what consciousness was, or
what the range of consciousness could be, on-stage or offstage. Thinking in terms
of categories, accidents, things, or substance, man is an intelligence, a soul, a spirit,
a mind, a psyche...united, more or less, accidentally or essentially, with a material
body. How can we be certain that we are really conscious of what we think we are
conscious of, and not merely oneirically imagining the whole thing? It is one thing
to analyze brain processes, but quite another to issue a deterministic statement on

its constituted function. Quite possibly, consciousness is a factor of "noise" in a system of discontinuous neural activities.

In the theatre, the perturbations we experience as thinking processes that mimic the fictive world of the stage characters may be as close a determination as we can make. In that case, complex adaptive systems can take into account this unpredictable, apparently random behavior of billions of neurons, firing across millions of synapses, and state categorically: "This brain activity is consciousness in action; this is what the actor-as-character is experiencing as I listen to and watch the stage action."

The link to theatre in neuroscience, biological, and evolutionary research in the last two decades is considerable. Similarly, the considerable link to indexical components of fictive stage space, which has always been there, has been largely unrecognized. In the art world, through the nature of the sign, all of these elements form a mosaic for performance and for critical analysis. For this reason, we can probably safely discard most intellectual approaches to theatre production as "interestingly irrelevant."

The *cerebral cortex* is approximately 2 centimeters thick, with a total surface area that, if stretched out to flatten the folds, would be the size of an average office desk. Neurons are polarized cells that receive signals on highly branched extensions of their bodies called dendrites, and send information along unbranched extensions called axons.

Most thought and perception takes place as nerve impulses, called "action potentials," move across and through the *cortex*. Action potentials are limited to about 200 per second for any given cell; travelling at a maximum speed of 10 meters per second, they must be recharged along the way to release their product, the "fleeting thoughts" of everyday life. The average thought of any one of us consists of thousands of connections made by each neuron, boosted along a pathway that is triggered by sequential timing of axons, firing in parallel, to make the necessary connections.

From the inception of life, the initial stages of axonal growth show a remarkably complex wiring diagram during embryonic life. But as Fichback notes, various kinds of axons, on arriving at a particular site, seem to be influenced by nerve impulses originating within the brain, or stimulated by events in the world itself. Thereafter, "synapse formation during a critical period of development may depend on the type of competition between axons in which those that are activated appropriately, are formed."[13]

To begin to understand the complexity of brain processes, and the reasons for discarding the outmoded methodology of Newtonian "linear" solutions, Gerald Edelman describes his theory of the mind of man as a dynamic process of recategorizations: the nervous system in each individual operates as a selective system resembling natural selection in evolution, but operating by different mechanisms. In

Neural Darwinism, Edelman makes persuasive arguments for the natural selection of neural groups and the creation of "neural maps" in the cerebral cortex as the basic synthesis of uniquely individuated man. Putting aside larger questions of whether evolutionary Darwinism or "punctuated equilibrium" (Stephen Jay Gould, R. C. Lewontin) is the correct ideology for the history of species on this planet (Edelman clearly in the Darwinian camp), the question Edelman asks is: "How does the variable brain deal with an infinitely complex world—an 'unlabeled world'—to make sense of things?"[14]

In terms of theatre and complexity issues, how does the playwright choose a text, deconstruct that text, and present to the spectator this theatrical amalgam, combined with a similarly deconstructed visual and musical book? How does the spectator synthesize the opposition between mental representations and reality, and "uncover" the performance? Leaving aside for the moment problems of theatrical displacement, Edelman's explanation involves two concepts: degenerate repertoires and cell adhesion molecules (CAMs).

In Edelman's view, the human brain operates by strengthening certain combinations of connections between cells in particular groups. Since there are more than fifty billion cells and a thousand trillion synapses, the possibilities for synaptic connections are virtually endless. Cell adhesion molecules function and vary during the development of basic patterns and connections. After these connections are made, conscious "Darwinian" selection occurs by means of re-entrant signaling, which implies that there are overlapping repertoires of neural populations that can be discarded as other repertoires prove to be more useful. These latter degenerate repertoires can then be converted into new positive re-entrant signaling repertoires. Edelman explains the function: "The part of the human brain that sees red must communicate with the part of the brain that sees edges before it can even begin to conclude that what it is seeing is an apple."[15] The mention of the signifier "apple" starts a re-entry process, whose signifieds include *redness* and *roundness*—and many more neural mapping connections. But in addition to this basic semiotic situation, in a theatrical situation a transformational "seeing" occurs, the semiotics of which began with the explosive power of a theatrical image.

Look at the apple as an image of life, of original sin, of sweetness. Look at the scene in Wilson's *The Forest*, in which The Whore takes Enkidu's virginity in a seven-day embrace, Edelman's semiotic neural structure might function as follows:

> The signifier "approach image" has a re-entry signified, with a referential point within the *limbic system*. Since the basic *paleo-mammalian* response pattern consists of one of the following: "fight, flight, hunger satisfaction, or sexual encounter," Enkidu's response is surely one of "sexual encounter" of the most basic kind.

With this semiotic neural structure taking place in an artificial, implicative theatrical setting, there is a theatrical displacement of conventional reality, less conscious, sub-subjective and real, that emboldens the dramatic beat with its own rhythms and referential systems.[16]

At this moment, primary (*proto-mammalian* reptile) significations merge with secondary (neo-mammalian, implicative, symbol system) significations to form potential concepts in the mind (the *cerebral cortex*) of the spectator. Referring to recent research cited above, evolutionary triggers embedded in the species from *Hadrocodium wui* (Greek words for "large headed") merge into 21st century consciousness by means of neural maps that connect the *frontal lobes* to the *pre-limbic amygdala*. These potential concepts are then realized as sound-images in *secondary association* areas of the *angular gyrus* in the left *temporal lobe*, known as human speech, sub-verbal thought, "flashes" of insight; or as deep emotions such as fear, dread, love, rage, and desire.

In combining this vertical semiotic with Jakobson's "transformed poeticity" and relating the mirror image "congress" of Mukarovsky's aesthetic object in collective consciousness to all the elements of stage production, a multi-form theatre aesthetic can be identified. However, it must be emphasized that no single paradigm will explain the vagaries of a situation except as a series of on-going dynamic relations between immensely complex vertical and horizontal modes of significations. Another "dualism in science" between human culture and prehuman nature has been replaced since the 17th century, in the words of Alexander Argyros, by a new view of the universe as a dynamic system.[17]

Argyros focuses on the long history of the emergence of consciousness, that involves the extinction of the reptiles prior to 65 million years ago, and the filling of the vacated evolutionary niches by diurnal "big brained" mammals, our ancestors (weighing in at 2 grams, about the weight of a paper clip.) Since the typical mammalian distance sensing system, with capacities for storing information over time, was an encephalized auditory and olfactory system, the newly visual species of mammals also developed a neuro-sensory system.

In an early text by Harry Jerison, the argument is made that this brain expansion is the foundations of consciousness.[18] The evolution of hominids to allow construction of perceptual worlds, such as we know today, "where it [is] possible to evoke a visual image by the use of touch and sound or a tactile or an auditory image by the use of other modalities such as vision or even odors," also lays out the capacities for reflection, for imagery, for language, and for culture.[19] Unfortunately, hominids only take us back 6 million years; we may have to go back 195 million years and start forward to the evolution of the full cerebral cortex. However, by providing a fundamental neural basis for categorization of the things of this world, as described by Jerison, Edelman's selective recategorization unifies perception, action, and learning.[20]

This particular openness and adaptability to incoming sensory modalities, which we also know as the discontinuous basis for language, allows the development of a theatrical sense in man, as in no other species. We have theatre because we evolved as we did, from nocturnal animals with a highly adaptive brain, into creatures with the capacity to fantasize and to envision from sensory inputs a world of meaningful associations that we call "the external world." Given the fuller description of man's ancestors, a description of *Homo sapiens* as the biologically-ordained theatrical species is almost tautological.

As regards to memory, and its place in theatrical fare, Edelman claims that we do not store images, or bits, but become more highly endowed with the capacity to categorize in connected ways. Neural maps are the key to understanding brain function, in Edelman's view, and there must be incessant reference back and forth among them for categorization to occur. And by referring the more abstract neural mappings back to the primary sensory areas, for example, which have a continuous relationship with external stimuli, the brain can effectively keep track of its various regroupings of the sensory inputs.[21]

Once again, there is no link made by Edelman and his co-workers between the complexities of orchestrating neural maps, and sequential differentiations that are in a constant state of flux, which can only be explained in terms of chaotics. But it would not seem unreasonable to conclude that the variable brain deals with an infinitely complex world, an unlabeled world, according to its neuronal maps. Following Jerison, Edelman concludes that arrangements of cells, *unlabeled in their usage*, are found in most brain regions, including those for vision and touch. Mapping that begins before birth continues through life, based on reentrant signaling between other neural populations and external stimuli. Eventually, in the twinkle of an eye, a thought is formed, as the deep structure of evolutionary biology rises above the almost uncountable hordes of neural maps that fire their impulses across thousands of synapses.

As for the principles of memory involved in this theory, Edelman claims that memory consists of an interplay between local cortical sensory maps and local motor maps: these, together with *thalamic nuclei, basal ganglia*, and the *cerebellum*, interact to form the global mappings that permit the definition of objects as a result of continual motor activity.[22] What is missing is the impetus that creates these circumstance: the *biological intentionality* that permits the mind to focus or to wander, at will. The presumed answer has always lain in the *cerebral cortex*, as the sight of "mind power," and new evidence shows the extent of these resources.

Elkhonon Goldberg suggests there is a progression across evolutionary time of controlling structures in the human brain. "Subcortical structures developed before the cortex,"[23] he remarks, "and for millions of years they guided the complex behaviors of various organisms." The *thalamus*, controlling the reception and processing of external stimuli, and the *basal ganglia*, in charge of responses, motor activity and behaviors, preceded the late evolution of the *cerebral cortex*.

When one looks at living reptiles and birds, there is only marginal *necortex*. When one looks at humans, the *neocortex* prevails and in particular, the *frontal lobes* of the *neocortex* dominate the brain landscape. For Goldberg, this dominance of the *frontal lobes* of the *neocortex* is the vital clue to consciousness. Following on the work of Hughlings Jackson in the late 19th century, Goldberg speaks forcefully of the properties of the *prefrontal cortex* as the "critical prerequisite of consciousness, the 'inner perception.'"[24]

The *prefrontal lobes* are the best-connected part of the brain, including the posterior *association cortex* (for visualization), the *premotor cortex,* the *basal ganglia*, the *cerebellum* (for motor movement and coordination), the *hippocampus* (for memory), the *cingulate cortex* (for emotion), the *amygdala* (for civilized behaviors), the *hypothalamus* (for homeostatic functions), and to the *brain stem nuclei* (for activation and arousal.) It would seem that intentional acts of the species are inseparable from *prefrontal cortex* control, whether we speak of *right frontal lobe* as the innovative lobe and the *left frontal lobe* as the regulatory lobe.

DISCONTINUOUS CHOICE WITH A STACKED DECK

When we do theatre, we replay the history of the species that extends far beyond the emergence of consciousness. When we examine the range of species since the Cambrian Age, five hundred millions years ago, there is the probability of 17 trillion kinds of creatures. Nothing like that number ever inhabited the earth. The record as it stands today, according to Jay Gould, shows too few vertebrates, no invertebrates after the vertebrates emerged, and the odds that, if we replayed the tape of species on earth, the chances of getting *Homo sapiens*—bipedal, two-eyed, conscious creatures—are practically nil.[25]

We can define "two-eyed," and "bipedal." But when we attempt to define consciousness, we run into a key problem that has eluded even the finest of our playwrights and performers. How do the physical events occurring in our brains, while we think and act, relate to our subjective sensations? —that is, how does the brain relate to the mind and to a unitary consciousness, which is what we want to portray onstage in any given moment? We may discover that the old definitions of consciousness are only a tiny portion of the real matter. Despite some points of contention, it can be speculated that, from a neuro-anatomical perspective, the human brain went through three phases before it became the modern functioning unit we describe as the mind of contemporary man. There may well be a fourth phase under way in the present era.

In the proto-reptilian brain, the *brain stem* and *reticular formation* form the fundamental core of the nervous system, composed of the upper spinal cord, parts of the *mid brain*, diencephalon and the *basal ganglia*. Behaviors are limited to the establishment of home territories and the finding of food and shelter and breeding.

Later, as the proto-reptilian brain developed, the *limbic system* evolved as Nature's first step toward providing self-awareness, and especially an awareness of internal conditions of the body. In this proto-mammalian visceral brain, the *hippocampus* provided a mixture of information on the inner and outer worlds of the creature and the *septal areas* provided internal information. Data on the world without was now provided by sensory systems. In the final neo-mammalian brain stage, the *neo-cortical mantle* developed. Responsible for cold (non-emotional) and fine grain analysis of the external environment, and the generation of expectancies about the future, creatures with this neural pattern possessed extraordinary advantages over all other species.

The first direct evidence of this heightened neural activity in art is found in Southern France and the Spanish Pyrenees. If perception is understood as the capacity of man to measure time present, to plan for the future, and to comprehend a sequential past, examination of the cave paintings at Lascaux, France, reveals striking information of the dawn of perception. Lascaux drawings include a stick figure hunter being attacked by a dying bison, fully developed with dripping entrails, a suggestion that the reality of the scene was very subjective. The power of death embodied by the bison has apparently overwhelmed the stick-image of the human figure.

The interpretation of the scene demanded and still demands the viewer's participation to experience that supreme moment of crisis, not as an external, but as an interfigural perception, in which objectivity plays no part. There is an implicit demand here for the viewer's consciousness of particular emotions—that of fear and flight. (The human *limbic system* has only a repertoire of four: fight, flight, sexual encounter and hunger.) To achieve a sense of power, most great works of art focus on one or more of these typologies.

NEURAL CORTEX IN CONTEMPORARY THEATRE

In theatre, biological intentionality or more precisely, *neural theatricalism*, is the concept that, in the elicitation and comprehension of performance there is a displacement, by definition less conscious, sub-subjective and real that predates presence. Somewhat surprisingly—and without explanation—in the late twentieth century, commentators and playwrights chose as their subject this sense of displacement in theatre *that has always been there.*

Through the 1970s and 1980s, Wilson in particular developed a vital new methodology that suggested a theatre taking place in the interstices between the conscious and the unconscious.[26] His 1985 production of a section of *CIVIL warS: A Tree is Best Measured When it is Down,* was a triumph of beautiful oneiric landscapes, elaborately calibrated lighting plots, and multi-scrimmed stage settings.

Wilson's 1986 American Repertory Theatre production in Cambridge, Massachusetts, of *Alcestis* did not hold the spectator's attention because he diffused the mythic possibilities of the text into a minimalist narrative. However, in his 1988 Berlin and Brooklyn Academy of Music productions of *The Forest*, Wilson regained some measure of the spellbinding images of earlier spectacles. The play also examines fragments of a biological intentionality that lie beneath conscious awareness. (There is a hint of Jaynes' "half god" right hemisphere voice versus the "half man" human of the left temporal lobe as the root of this Middle Eastern tale.)

Based on the ancient Babylonian epic of *Gilgamesh* (half god, half man) and (half man, half-animal), Wilson's commentary reveals his thinking on the relation of the text to playwright, actor, and spectator.[27] Adopting artistic deconstruction as a major statement of his project, Wilson's "Director's Notes" to the audience are very explicit:[28]

On the playwright-as-image-maker:
Artists are people who find new languages, and destroy those languages, and make other languages out of the destruction. That's what choreographers do, composers, writers.

On dealing with actors::
You could call my way of directing anti-Stanislavski.... To me when Howie [Seago—ENKIDU] goes to put his hand out to the woman, he doesn't know why he's doing it. If I were ever to think of an idea to tell him, it would just be destroyed, because it's so complex.... His gesture has everything, and so we can't limit it.

On rehearsal technique:
I felt that the first text was important and that from it we would make the language of the piece. By destroying it, and then putting it back together, then destroying that.

On audience focus:
[In my theatre] the eye can see one thing and we can hear another, as happens all the time. It's through this paradigm that one thing reinforces another... If we don't illustrate in the audio book the visual imagery of the text, then the visual book of the audio screen is boundless. And if we don't illustrate the visual book with the audio book, then the visual screen is boundless. The problem with most theatre is that it gets boxed in.

On actor training::
When Medea goes to the children: "What is this?" And maybe that's why Medea is still modern. Because we are all capable of that murder. That's why I can't tell Geno [Lehrer—THE WHORE] what the situation is when she's walking to Howie [ENKIDU.] There's no way I can tell her. It's too complex. And she's a great artist, and she's great when she's doing it.

On communication:
The authentic experience in the theatre is the uncovering. Theatre is not something we can comprehend; it's something we can experience. What's necessary in theatre is the audience. What happens in the audience is the mixing.

Wilson is also good at telling visual jokes in which the audio book is used to explain the visual imagery. A short "Leg Piece" can serve as an example: A man stands on stage beside a square of light; offstage the sound of an elevator begins. Sound of elevator door opens. Man steps into the light square. Sound of elevator door closing and then sound of the lift engine begins. The whining increases, stops. The sound of an elevator door opening. The man steps out of the light. He walks to another space. The square of light comes up beside the space. Sound of an elevator begins, and the sequence repeats. What has happened? The man "has moved to an-other floor" in the visual book, the conceptual experience, and the audio book. A displaced sound-image is reconstituted.

PARADIGM SHIFTS

The more conscious brain is far more conventional. It's good for getting to the store on time and stuff. But you have to get rid of that voice or subdue it on some level.

Neil Simon

So at that moment I'm the person sitting there in the chair, thinking, "I hear this voice, I know that's somebody that I know...." And then once you get it "voiced,"...the writing can begin.

Marsha Norman

In 1970, in *The Structure of Scientific Revolutions*, Thomas S. Kuhn described the discontinuous advancement of knowledge by the sequential development of what he termed a revolutionary paradigm shift.[29] This shift occurs when a prevailing system of thought can no longer explain the anomalies of a theory. Ptolemaic astronomy, for example, was replaced in 1543 by Copernicus's revolutionary paradigm on the placement of the sun in the cosmos. In 1985, two theatre paradigms were proposed. Joseph Roach's *The Player's Passion* examined the science of acting, and suggested that Stanislavski's teachings and the proponents of American method acting, had deflected research away from competing theories, in particular Diderot's *Paradoxe sur le comédien* (1773). Subsequent acting theories and commentary, in many ways derivative of Stanislavski, have not clarified an already mysterious process. As Roach noted:

These terms and concepts do not add up to anything approaching complete answers to the problems posed by the actor's art, then or now. That we still regard the creative process and the performance event as miracles of a sort is credit to their fugitive nature and to the fascination that their contradictions continue to exert upon us.[30]

Roach examined actor's passions scientifically, biologically, experientially, over the course of Western civilization, using Diderot's *Paradoxe* as a new paradigm. He concluded that since the eighteenth century, physiology and psychology have sought to demystify the actor's "act of revelation." In consequence, the act of revelation that elevates the true actor's feelings into a kind of fiction—a work of art if done well—is further from being understood today, in the age of Stanislavski and the proponents of his scientific approach to acting, than it was in Diderot's.[31]

On a personal note, during a visit to Sam Beckett in 1985, the playwright told me a story of his friend, Jack Yeats, who he had not seen since before the Second

World War. On returning to Ireland after the war, Beckett had visited with the great Irish painter and asked him how and in what way had his paintings changed over the years. After a long pause, Jack replied: "Less conscious." For Beckett, on the verge of his own explosive decade of creativity, the words were like lightning bolts. Within three years Beckett composed his Trilogy of novels, and saw *Waiting for Godot* into production.

Recent research in neurology suggests that the ideology of literary and dramatic criticism and of the human sciences, anchored in nineteenth century paradigms, may be on extremely shaky ground. The first crack in the positivist-cum-postpositivist edifice came in 1986, when British anthropologist Victor Turner, revered mentor to Richard Schechner's "Performance Theories," renounced hallowed axioms of his own, of his generation, and of several subsequent generations; axioms that expressed the belief "that all human behavior is the result of social conditioning."[32] Taking a page from both Foucault *and* MacLean, Turner discussed the geology of the human brain, noting that each strata is alive, and that "even our proto-reptilian and paleo-mammalian brains are human, linked in infinitely complex ways to the conditionable upper [neo-mammalian] brain and kindling it with their powers."[33]

Chapter Three

◦⟨❀⟩◦

Interstices

TALK ABOUT INTERSTICES!

In the 1980s, the position of postpositivists and phenomenalists who called for a relocation of the theatrical event "at some mid-point...of intersubjectivity between performer and audience" (Favorini), or who defined theatre as a fundamental mode of relationships that intervened between live actors and audiences (States), or who described "mimetic fusion" between actor and spectator (Wilshire), accentuated Foucault's 1972 emphasis on the play of *différence*.

This relocation was not original to Foucault. In 1934, Jan Mukarovsky declared "the autonomous existence and essential dynamism of artistic structure."[1] The work of art—in our case, the theatrical event—"is identified neither with the individual state of consciousness of its creator nor with any such states in its perceiver nor with the work as artifact," but bears the character of a sign. Declaring the art work to be an aesthetic object located in the collective unconscious, with the artifact functioning as a perceivable signifier, Mukarovsky rejected his contemporaries' view that the art work was "a direct reflection of the psychological or even physiological states of its creator or direct reflection of the distinct reality conveyed by the work...."[2] By denying the validity of any subjective mental state and any hedonistic theory of aesthetics, Mukarovsky and adherents to the Prague School neutralized the creative process in favor of the creation of a constant dialectical relationship with an outward signifier and a collective unconscious.

Directly related to this issue, neuroscience of the 1930s was emphatically not the neuroscience of the 1980s and 1990s. But the Prague Linguistic Circle began with a prior scientific agenda that had more to do with Jungian archetypes than dramatic art—in this case; art in general for humanity's information. Furthermore, as one commentator has noted, this semiotic point of view valorized the terms "art" and "aesthetic," giving them a hallowed, metaphysical status that rendered them useless except as ikons.[3]

Relationships of groups of similar elements—sets of apples to sets of oranges—rather then direct comparisons, described the methodological approach of the cultural archeologist, the postpositivist, and the phenomenologist in the theatre. Even Richard Schechner, summarily announcing the imminent demise of drama-

turgy and of conventional theatre as the detritus of a Eurocentric culture—at the
very least, an English ethno-centric bias we ought to be reaching beyond—
suggested that ritual performance occupied a space "between the performer who is
doing the action and the spectator who is receiving it."[4]

The basis of the New Critics' analytical dichotomy—does the text mean what
the author wants it to mean or does the text mean what the speech community at
large takes it to mean?—had been consigned to the literary dustheap as naïvely irrelevant. Radical subjectivity, the basis of Modernism, derived principally from the
works of Anton Chekhov, had no counterpart in the critical world.

As noted earlier, Hirsch was acutely aware of the subjectivity and relativism of
contemporary New Critic commentators who, he claimed, "destroy[ed] any basis
both for any agreement among readers and for any objective study whatever."[5] However, Hirsch's critique of Wimsatt and Beardsley's "Intentional Fallacy" essay was
itself the best reason to abandon positivism. Hirsch made a point of singling out
Rene Wellek and Austin Warren's doctrine that the textual meaning of a document,
leading a life of its own, changes over the course of time.[6] The question Hirsch
posed, "Is it proper to make meaning dependent upon the reader's own cultural giv-
ens?" has been soundly answered in the affirmative.

Further, Hirsch's apocalyptic protest against the concepts of the New Critics:
"As soon as the reader's outlook is permitted to determine what the text means, we
have not simply a changing meaning but quite possibly as many meanings as readers," was exactly the point of Wellek and Warren's "perspectivism." As Wellek remarked, perspectivism, holding that values "grow out of the historical process of
valuation," required that the historian "refer a work of art to the values of its own
time and of all the periods subsequent to its own."[7] The resulting tensions of ambiguity, as Welleck and Warren described them, and the postpositivists' belief in the
multiplicity of paradigmatic readings of relations to a text, are part and parcel of a
realization that objectivity is largely an illusion. This rejection was perfectly in
keeping with Hirsch's own doctrine of objectivity. In a real sense, we have simply
broadened the "horizons" (Hirsch's term) of norms and limits that bound the archeology of meanings represented by the text, or script.

In his essay, Hirsch's conclusion—that interpretation "is the construction of
another's meaning"—does not vitiate the intentional fallacy arguments since *another's* meaning might well be, in the context developed here, the playwright's, the
actor's, the spectator's biological (conscious and unconscious) intentions. The old
paradigms of "consciousness" and "the unconscious," so popularized by psychoanalytic literary theory, simply do not stand up to recent neurobiological research.

There is not space here to elaborate on the neurophysical dynamics of "perspectivism," beyond indicating, as I have in the foregoing, the biological constructs of
Geschwind and Edelman and their implications for further research in aesthetic
theory. But it is germane to examine certain artists' intuitive sense of this neural

perspectivism; in particular, a discussion of commentary by those playwrights who are on record as acknowledging a "less conscious" process in the art of creating *sub-intentionally* (in the normal sense of the word) their works for the theatre.

MACLEAN UNLEASHED

In 1970, neurologist Paul MacLean proposed that man had a triune brain.[8] Neural and behavioral distinctions can be made, according to MacLean, among three types of systems: the *proto-reptilian brain*, representing the fundamental core of the nervous system; the *paleo-mammalian brain*, composed of a *limbic* system (contributing to man's first steps toward self-awareness); and the *neo-mammalian brain*, providing "fine grain" analysis of the external environment.[9]

In MacLean's model, the notion of subjectivity takes on new meaning. There is now a suggestion of a hierarchical brain structure relating to behavior. For MacLean, the *limbic* system (the *paleo-mammalian brain*) at the center of the human brain, is able to govern the ancestral constituents of behavior found in the *proto-reptilian brain*, and to provide "a necessary component of, or context for, significant memories" with which the *cerebral cortex* reacts to the external world.[10] In other words, the *limbic* system is the vital link between raw instinct and considered response.

Since the endowment of memory function is seen as a mid-brain process, beneath conscious control, this concept has great interest for theatre aesthetics. Using MacLean's schematic, the development of language (its sound-image components centered in the *angular gyrus* of man's left *temporal lobe*) is a product of the biogenetically recent *cerebral cortex*; perhaps no older than 35 to 50 million years. Since the core of the theatrical process (the pulling of emotional materials from the *proto-reptilian brain* and the re-siting of memories from the limbic system in the *paleo-mammalian brain*) lies much deeper than *neocortex* functions, conscious intellect can now be thought of as a stringed puppet, a fine-tuned marionette, with the marionettes in charge.

NEURAL TYPOLOGIES

In recent decades, theatre historians, following the lead of Michel Foucault, were more fortunate than theatre theorists. In 1972, Foucault fumed that traditionally the document was "treated as the language of a voice since reduced to silence."[11] Historians of the past viewed documents as inert objects that could be reconstituted to reveal what men had said or done. "Not so," stated Foucault, the goal was not to interpret a document but to work on "the interplay of relations within it and outside it."[12] This was an archeology of social strata that consisted of intrinsic descriptions of documentary materials as relations of unities, series, or tables.

In the 1980s, a postpositivist methodology emerged that resembled cultural archeology. The cultural archeological historian's newly described job was to define these series' limits, relations, strata, etc., over long periods of time; to individualize different strata; and to juxtapose documents from widely spaced periods without reducing them to linear schema.[13] It is tempting to suggest that dramatic theorists consider moving beyond the semiotic geography and deconstructive geology of theory-driven, abstract symbols of reality to embrace equally the emotional, somasthetic, preconscious elements of communication in the theatre—the origins of Peter Brook's holy and rough theatre—of which images, language, gesture and comprehension are major intellectual products of sub-strata *displaced* along the way.

DISCONTINUOUS PROCESSES IN THE VISUAL ARTS

Assembling the collected evidence of seventy years of theatre aesthetics, the organization of *neural theatricalism* might be stated as follows:

> Re-entrant neural mapping processes of degenerate repertoires of the cerebral cortex, triggered by proto-reptilian and paleo-mammalian behavior functions, intentionally *displaced* from their unconstituted and divisible origins, are the essential ingredients of performance.
>
> Conditioned by knowledge of the special transformational nature of dramatic performance (of the intentionally *transformed* sign composed of a signifying artifact created by an artist, an aesthetic object registered in the collective consciousness, and a relationship between the artifact and the aesthetic object) and giving visible body to the Other that was never there, interpretation of a performance is possible.[14]

There is at least one further crucial ingredient involved in successful performance: the capacity of the artist to produce dynamic displacements, transformable concepts that press the boundaries of unconscious relations where none were thought possible. This articulation can best be understood by looking at the pictorial images of Pablo Picasso's *Les Demoiselles d' Avignon,* and the works of the Cubist period that immediately followed.[15] His fascination with tribal masks was not so much with works of art but with ritualistic objects, "weapons for exorcism," that he could tap to reveal the resources of the art of the past.

In his review of the Museum of Modern Art's 1990 Picasso-Braque exhibit, John Golding suggested that Picasso's Cubist images "[were] themselves and yet simultaneously [could] evoke analogies with other things:...women's bodies [were] like guitars or violins, touchable and strokeable; dots or pegs [could] represent eyes or nipples or navels."[16] The demonic frenzy of these displacements at a sub-verbal level, carried to the point where Picasso (and Braque) "ended by making the very process of image formation virtually the subject of their pictures," formed an exact parallel seventy years later with Robert Wilson's theatre of images.[17]

In *Les Demoiselles,* it is not that two of five women have become African masks, but the implications of the boundaries of these new relations of women, relations of primal power—of grave, apocalyptic, less conscious menacing power, of *différence* from Cézanne bathers—to the viewer. In the theatre, the viewer response might be similar. By suggesting *mimesis* as a basis of theatre, Aristotle was not wrong but only half right: not "imitation of an action," but *displacement of an image* is the essential creative act, from which cognitive acts begin. What has been said of the image could also be said of the playwright's text, and the actor's response to that text.

It is certain that the actor's truth is not machine-like, but involves in part the development and continual recategorization of interacting neural maps that become fixed during rehearsals, or from years of study.[18] The development of technique ensures that each performance, monitored by continual reference to primary sensory modalities, has consistency. Social, anthropological and even political theories of the study of drama do not, in this "beyond deconstructionist" view, explicate theatre. Rather, these humanist theories demonstrate the secondary diversity of the mimetic urge, of which theatre is one primal expression.

In specific ways neural theatricalism implies that the playwright, the spectator, the director, and the actor experience and re-experience traces of flow structures of nerve impulses among well-defined cell groups. The truth of these intentional recategorizations (by the artist), or sympathetic recategorizations (by the spectator), are by definition arbitrary, differential, and discontinuous. The question remains as to whether the performer represents, in the spectator's mind's eye, a never-to-be-realized primal substitution (as Blau would describe the situation), or a stimulus for the neural processes to begin.

In *The Forest*, as in his other folk operas, Robert Wilson invites spectators to find in his stage images primal substitutions for new relations, one image to another or one series of images to the text of another, at the very boundaries of what is conceivable in external reality and what-can-never-be, or to "colonize" re-entry neural mappings of degenerate repertoires of a virtual reality not-yet-experienced but no less real than the traces of a triune brain that filter through the interstices of the conscious mind. Viewed from the perspective of complexity studies, Wilson introduces the spectator to two self-contained and self-perpetuating worlds—carefully differentiated—and then invites any chaos factors of the spectator's choosing to spin out mandelbrots of meaning/s. The results can be comical, illuminating, diabolical—or for those who do not become engaged—extremely boring.

The speaking subject of Samuel Beckett's *The Unnamable* also pinpoints the nature of the creative act. Opening on the I of *The Unnamable*, where "I" resides, Beckett's threshold door to silence is a continuing metaphor of the nature of performance as displacement of a primal substitution. I say "performance" because this first person novel was written immediately after *En Attendant Godot,* which Beckett wrote in 1948 "to put some limits on the darkness."[19] Both playwrights—Beckett and Wilson— work the same neurobiological terrain.

Chapter Four

·❦·

Chaotics

CHAOS AND COMPLEXITY

James Glick's *Chaos: Making a New Science (1987),* was the early marker of Chaos Theory. The book describes three radical transformations in twentieth-century science: Relativity, which eliminated the Newtonian illusion of absolute space and time; Quantum Theory, which eliminated the Newtonian dream of a controllable measurement process; and Chaos Theory, which eliminated the Laplacian fantasy of deterministic predictability.[1]

Chaos theory applies to much more than weather patterns, of course, and systems that account for physical phenomena in the external world seem to play no less a part in the internal, biological world. As Glick further notes,

> At the pinnacle of complicated dynamics are processes of biological evolution or thought processes. Intuitively there seems a clear sense in which these ultimately complicated systems are generating information. Billions of years ago there were just blobs of protoplasm; now billions of years later here we are.[2]

With 100 billion neuron cells in the human brain, an explanation of the mind of man as a expression of a continuous linear pattern of performance along the line of Newtonian mathematical reasoning, or Descartes' division of mind and body, is hopelessly inadequate. Chaos theory can be simply described as the ultimate wave effect. The body movements of someone practicing tai chi in Beijing Square will affect the weather patterns in New England, and so on. More important to the study of theatre and theatrical forms is the notion of fractal zones or basal boundaries, which is primarily the study of systems that could reach one of several non-chaotic final states. To situate an image or a phrase on the boundaries of several interpretations—to "vaguen" the image—so the viewer/listener must work the consequences of the situation within neural maps of the cerebral cortex, is obviously a goal of any artist. Unspoken as it has always been, but constituting one of the mysteries of "art" for all time, these fractal zones of sound and image potentials, are further complicated by the deep consequences of biological fractal zones that can also be used in a performance dynamic. The Mandelbrot fractal curve implies an organizing structure that lies hidden among the complications of such shapes. Everything from the beating of the heart to the patterned sequences of neural firing in the hippocampus, to facilitate learning and memory, has fractal implications.

If Aristotle was correct, and theatre gives pleasure because it is a learning experience, then it belongs at the heart of chaos theory, not the chaos of disorganization, or the chaos of "theatre and war," but the center of an evoked chaos of the brain, where competing neural systems create images and intellectual concepts that allow communication to take place, either of a receptive or a forged kind. To this can be added the certainty now that it is not a single brain but many brains or many neural systems in the brain, that lead to the formation of "mind," as a single entity. That three-pound organ is the most complex structure in the known universe. According to Gerald Fishcake, the human brain, "with imperatives in the *limbic system* and geographic information in the *hippocampus*" can, as a result of millions of years of evolution, "create a three-dimensional landscape from light that falls on a two-dimensional retina."[3] With a estimated 12 billion neurons, each capable of forming as many as 50,000 connections with other cells, Kathleen McAuliffe estimated that 100 trillion connections are possible—theoretically.[4]

Everything that we produce onstage is fine-tuned through the cerebral hemispheres, a cell-rich, laminated cortex two millimeters in thickness, with divisions between numerous sensory receiving areas, motor-controlled areas and less well-defined areas in which associative events take place. For Fischback, it is here in the interface between input and output, that the grand synthesis of mental life must occur.[5] In strong agreement, Glick remarked that "even when a dynamical system's long term behavior is not chaotic, chaos can appear at the boundaries between one kind of steady behavior and another"[6]. Mind is not a single entity but a sequence of processes, a grand array of fractal possibilities, the unfolding of information in milli-seconds, and synthesized into the unitary consciousness of mind. The seamless presentation of a character or a scene in a stage production is nothing compared to the seamless presentation in the healthy *Homo sapiens* of the appearance of unitary consciousness that must underlie comprehension, learning or communication. The consequences of the failure of unitary wholeness is, as we well know, schizophrenia or mania.

As for the perfectly clear and lucid moments of inspiration, something approaching those few days that Nietzsche experienced in 1888, consider the commentary on fractal boundaries, "in the midst [of which] may come the picture of stability"—the absolute expectation of every aspect—the Mandelbrot sets brought alive[7]—with every tendril and every stem in place—all signaling a universality of chaos and disorder as constituent of a deeper order. In moments of extraordinary lucidity, no one in the twentieth century expressed this sense of chaos and order in theatrical terms, or described theatre's mission as a recasting of the purity of life, better than the messianic Antonin Artaud. In analysis of texts from the perspective of the new theatrism, the job of the reviewer, the artist, the actor-designer-director in the twenty-first century, is to unfold that which has not yet been said, about the proposition that is presented onstage. The unfolding goes back to neural pathways in the brain, in the first instance, then to the history of neural creativity as it developed in the artist from the point of creation to the point of expression now, and then to the possibility of creative history in the species. In this small beginning, a range of possibilities exists, to carry the observer back into the very processes of life on earth. The possibility also exists to carry the artist forward, if the inspiration exists. At each way-point, the spectator/reader is brought to the realization that something is taking place, a portion only of which he or she can experience, both consciously and unconsciously, at any given moment.

The phenomenologist's parallel experience to theatrical creativity can be seen in the analysis of Merleau-Ponty, describing the painting of Cézanne's *Mont Saint-Victoire*. As Merleau-Ponty senses this most personal experience of the painting, Cézanne's landscape is as a "present and living reality": how we are open to this mountain, seen in this light from a perch above a farm across the plain.

The moment is not of one taking place in time and space, but of a making possible the spatiality of Cézanne's world in an expression of primary perceptual significance. All this is done "in a mode of comportment that concretely works out [the artist's] understanding, so that others also may experience what [he] has understood."[8] Rather than focussing on consciousness as a given, Ponty deals with the experience of participating with Cézanne as the artist created the images.

That experience of process transcends in important ways the assumptions of Newtonian physics in a belief that all can be measured on a rationally, "abstracted from nature scale," which led to the crises in art and science in the twentieth century. The twenty-first century will get it right, too, for a moment.

LIMITS OF DISSIPATIVE CHAOTIC STRUCTURES

One might summarize the conception of consciousness for much of the twentieth-century in the words of Virginia Scott, who wrote, in The Common Reader, that "Life is not a series of gig lamps symmetrically arranged; life is a luminous halo, a semi-transparent envelope surrounding us from the beginning of consciousness to the end."[9] Chaos theory has taken consciousness and opened it to the universe, where it might now be assumed that the envelope of reality takes us back to the Big Bang! All that came afterwards is our envelope of reality, from which we can not escape. Once set in motion, the laws that govern chaos and its fractal boundaries—in physics, in life, in theatre—have no point of outside reference. But even this postulate is a limited view.

At the beginning of the third millennium, scientists can speculate there was not one big bang, but many, continuing from the past into all of time and beyond. Our big bang is bounded by the speed of light, 186,000 miles per second, but why not another big bang, bounded by mega-photons of a billion miles per second, all contained within a folded mega-universe. What sort of creatures would exist in that universe and would they be any less real than those in our own.

A number of terms that relate to theatre and chaos need some amplification before Artaud's insightful writings can be examined. In the first place, the relationships between art and science have moved in tandem throughout history. Discoveries in science relate to the questions that society puts to its great thinkers, or leave implicit in their absence, such as Copernicus, and Machiavelli. As I have noted elsewhere, Princeton physicist John Wheeler pointed out that discoveries in quantum mechanic are intimately dependent on our perception of reality. There is, in other words, no such thing as "pure" science.

In 2001, Nobel Laureate Stephen Weisberg thinks we are moving steadily towards a "satisfying picture" of the world, where we will have achieved "an understanding of all the regularities that we see in nature."[10] Reality in the recent model of the 1990s involves many special interests and competing agendas, as Katherine Hayles noted, including "feminist critiques that explore gender issues, sociological studies that investigate how science is socially constructed, and ethnographic analyses of scientific communities as tribes with their own vocabularies, rituals, and social practices."[11] Much of Hayles' discussion is fervent conformation of Wheeler's commentary on self-interest as a dominant feature of any kind of research.

For the classical theorist, Isaac Newton's metaphysical model of the universe as a smoothly running clock underpinned modern science. This model had a number of planets not yet discovered, comets and other objects, and it was necessary for God to "push with his divine finger the reset button of the universe annually in order to keep things going as they were supposed to do."[12] Support for Newtonian metaphysics came from Laplace, who claimed for the nineteenth-century the dream that "the complete predictability of systems. based on Newton's laws of motion, could be achieved *so long as the initial conditions were known.*"(emphasis added.)[13] However, when nineteenth-century mathematician Henri Poincare attempted to apply Newton's "laws of motion" to three celestial bodies, the results were confusing. No single answer worked for three bodies, whereas results were always predictable for two. The linearity model was broken.

Since we know the very definition of Chaos theory is defined as "sensitive dependence on initial conditions," which can never be eliminated, Laplace's and Newton's unassailable places in modern physics were doomed. The classical paradigm had been broken. In specific terms, the difficulty of dealing with Darwin's perfectible species and implications of the second law of thermodynamics, led to the discovery of what Ping Chen described as "dissipative structures" and "deterministic chaos."

Much of physics regarding planetary motion was quite accurate. However, the mechanism of evolution had raised the most fundamental challenge to classical physics. As Chen describes it,

> The time arrow of the second law of classical thermodynamics is from order to disorder, while Darwin's theory indicates a general trend of evolution from simplicity to complexity. The attempts to resolve this contradiction led to the birth of no-equilibrium thermodynamics and the theory of self-organization, pioneered by Ilya Prigogine... In open system with energy flow, matter flow and information flow [i.e., in theatre performance], dissipative structures may emerge because of nonlinear interaction.[14]

These dissipative structures are the stuff of punctuated equilibrium and the basis of saltatory changes in biogenetic evolution that, in the case of *Homo sapiens neandertalensis*, led to the emergence of *Homo sapiens* in a few millennium, and the disappearance of Neandertal man almost overnight.

Early modern man possessed primitive communication skills, possibly rituals, essential to keeping the tribe together, that survived for hundreds of thousands of years. But creativity celebrates disequilibrium. More importantly, as Ilya Prigogine suggested, it is "not equilibrium but non-equilibrium that may be the source of order." Increased cranial and vocal capacities of *Homo sapiens* and adaptations to social and environmental demands brought increased capacities for quantum leaps of disorder. Since "nonlinearity interactions play an important role in emerging complexity and self-organization," the irreversibility factor ensures that a useful change will not revert to the old ways. This new paradigm of non-equilibrium and nonlinear science, or "self-organization" of *strange attractors*, represents a revolution in fundamental thinking.

Implicit support of this view of science comes again from Steven Weinberg's questions as to "why the genetic code is precisely what it is or why a comet happened to hit the earth 65 million years ago in just the place it did" or why there are deep mathematical theorems "that show the impossibility of proving that arithmetic is consistent." The most telling argument against classically-derived science is Weinberg's hope, in a century or two, that scientists will find a set of final laws of nature "rich enough…to allow for the existence of ourselves."[15]

Complexity Theory, begun in the 1940s in Santa Fe, New Mexico, in conjunction with Los Alamos' scientists, began to flourish in the late 1980s as the keystone to the next wave of natural science—not as an antidote but as a deeper model of the universe as we know it.

The term originated, according to George Cowan, in a 1948 article in *The American Scientist* by Warren Weaver. He predicted that "the study of complexity would become either a major concern of science or the major concern of science." Cowan concludes: "I'd give him credit for coining the terminology and predicting the growing importance of the subject."[16] Describing itself as "an institution that seeks to catalyze new collaborative, multidisciplinary research; to break down the barriers between the traditional disciplines, to spread its ideas and methodologies to other institutions; and to encourage the practical application of its results," SFI[17] continues to flourish in the twenty-first century.

Strange attractors is a descriptive phrase of a self-regulating system that combines principles of order and chaos. Within the system, there is a deep order, controlled by "equations" that regulate its behavior, and which are not visible or predictable to linear analysis. Behavior is totally unpredictable, but randomness is constrained because of the rules of the Chaos game.

The dynamics of the game—the theatrical performance, the behavior of a weather pattern, biological rhythms, are unpredictable, but constrained by the parameters within the margins of the attractor—the stage script, the range of temperatures on earth, the dancing heart beat, etc. In the field of theatre, Antonin Artaud was pushing chaos theory in the late 1920s, describing "plague" as the core of artistic creativity, analyzing in his own genius the implications of what he instinctively knew to be another way out of three centuries of Cartesian mind-body dualities. Artaud attempted to push the limits of comprehension, where fluctuations were a dominant mode, without benefit of modern chaos and complexity theory.

Within the eye of a complex system, a "bifurcation tree" will emerge and the system created may jump to one of its branches and reorder its chaotic patterns. Since this bifurcation process is irreversible, according to Chen, history enters into the bifurcation tree of a nonlinear dynamic system, and the world is changed.[18] Or so Artaud hoped! In exploring the areas between the plague and everyday reality, what Artaud was doing, in chaos theory, was to explore the fractal basin boundaries of competing systems. At the boundaries of each system chaos appeared, But within each system lay an unpredictable calm, a complexity, which may well be what Chen has concluded is to be the next great transition in our conception of the world: a marriage between occidental methodology and oriental wisdom.[19]

Beyond Katherine Hayles' new potential social and cultural hegemonies, challenging the traditional scientific communities, Alexander Argyros argues that the deconstruction revolution, "poised to resolve the seventeenth-century dualism between human culture and prehuman nature," offers no alternative "to the metaphysics of presence other than the affirmation of the continuous need to deconstruct it."[20] Derrida is not the light to be sought.

Rather than the paths of metaphysical closure and deconstructive demystification, Argyros argues that the most complex things in the universe, "including human beings and their theories, works of art, and social structures, are best understood...as chaotic systems." Equally important, Argyros notes anthropological evidence as integral to the processes of the modern human mind. The natural world's evolutionary track led the species to where it is today, and "by giving the natural world its due, ... we may he able to stop deconstuctionist's 'mad dash to ... disassemble anything that smacks of truth.'"[21]

His argument implicitly recognizes the triune brain of man, and the history of the development of the relationships of that divided brain's functions in creating performative acts on stage. As Irwin Schrodinger, the brilliant quantum physicist and father of quantum biology, reminded us so many decades ago, a living organism has the astonishing gift of concentrating a "stream of order" on itself and thus escaping the decay into atomic chaos."[22] Artaud fought for that sense of calm within; a revelation of those dissipative structures of the mind at the heart of his "theatre of blood"—as he described it in February 1948—all the days of his life.[23] Twelve days later he was found dead in his room, seated at the foot of his bed.

THE SMELL BRAIN

Another approach to the problem of discontinuity in performance might be described as "the New Theatrism," which seeks to engage the issues that underlie conflict as it arises on the stage in various dresses. Social, cultural and political crises pinpoint the sources in external reality, and then open the way to a deeper investigation of the nature of the species. Described by Keir Elam as "chaotics," these flash points can, from time to time, engage the viewer in social and cultural rediscoveries of the original acts now represented on the stage.

A proponent of this kind of investigation, following in the footsteps of Foucault, is Stephen Greenblatt, who describes the new historicist's job as a project "to uncover the moments at which works of art absorb and refashion social energy, en endless process of circulation and exchange." Looking in particular at Elizabethan plays, this English Department professor suggests that:

> The plays ... are shot through with borrowings, assimilated cultural traces ... [that] could and did have a real and important impact on society at large.[24]

These same traces are the basis of Performance Studies, the performative acts that serve as traces carried forward in the texts, and in the actions of performers, of cultural and social values from historic times. The problem is these traces are linear. Greenblatt's references move outward, but have nothing to do with the workings of the Mind, as Artaud described it, and "the fossil imprints so close to our own origins" as cited at the beginning of this text. As with Derrida, Foucault is also influenced by his Cartesian roots. Those cultural and biological traces that infuse every performance, from which we can trace the roots of our very selves in the workings of a quark or the burial customs of *Homo sapiens neandertalensis*, or the "smell brain" of a tiny forest creature some 65 plus million years ago—that preceded in a direct line our own species, are not limited to cerebral ramblings.

In short, the interpretations are facile, linear, cerebral, and not justifiable except as language games, of which deconstruction is the latest trace. The exclusion of the natural world, or the very face of nature of the actor, from any equation that attempt to interpret culture strictly by socio-historical pressures, is immodest, to say the least. The tiny animal—*Hadrocodium wei*—weighing in at about 2 grams, might well be our great, great grandfather, 195 million years removed. As Argyros argues, "the most robust and complex things in the universe, including human beings and their theories, works of art, and social structures, are best understood neither as Platonic essences nor as random processes, but as chaotic systems."[25]

The fact that the human species can associate two non-limbic stimuli gave modern man the power of reflection and of speech. But it was not the power to dismiss those regions where quantum leaps in biological capacities forever changed the evolutionary tape. Even in Elizabethan plays, perhaps more so than in many other critical eras, the dimensions of communication are a miracle of chaos theory that began in evolutionary history and progressed through dissipative systems of neural synapses. In the New Historicism/Theatrism, to look to the power of traces of social relations that appear to be implied in these works is to render the power of theatre itself as facile and one-dimensional, more suited for understanding in the library or in the seminar room, but not in the living, illusion-building laboratory of creation, participation, and interpretation.

Current "politically correct" issues can be cut through and laid bare, with all the pain and fury of the original long-smoldering debate revisited at any time. No need of traces now. Artaud, addressing fictive Chancellors of the European Universities, might well have had in mind Stephen Greenblatt and associates, when he announced:

> The fault lies with you, Chancellors, caught in the net of syllogisms. You manufacture engineers, magistrates, doctors, who know nothing of the true mysteries of the body or the cosmic laws of existence, false scholars, blind outside this world, philosophers who pretend to reconstruct the mind. The least act of spontaneous creation is a more complex and revealing world than any metaphysics.[26]

For the Chaos theorist, the determination to examine the processes of reality from anthropological, neurobiological, and from chaos theory perspectives, can bring both theatre history and critical interpretation to a common ground; the discontinuous processes that continually renew themselves as they unfold in dissipative structures that are played out within deterministic chaos of fractal boundaries, offer a rich new perspective. This is an endgame, to be sure, that continues *only as long as the process is contained* within boundaries of the original dissipative structures.

For the deconstructionist, the problem is one of irrelevant and inescapable objectivity: representation of traces is never outside the game of reference. Once begun, the game is unmasterable. The world is as experienced. In a similar light, the world of Artaud's theatre despised anything that smacked of the theatrical, of imposed forms, of culture-strewn artifacts, anything that detracted from raw nature, impaled on the stakes of societal probity.

COMPLEXITY ABYSS

While Artaud never met the wily old Nietzsche (1844–1900), the conceptual model of drama as the highest achievement of the species was common to both men. Artaud was only four years old when the great German philosopher died, having written in his last published work, "Man, the most courageous animal, and the most inured in trouble, does not deny suffering *per se* ... he seeks it out, provided that it can be given a meaning." And this seeking is "a will to nothingness, a revulsion from life, a rebellion against the principal conditions of living." The irony, for Nietzsche, was that man "would sooner have this void for his purpose than be void of purpose."[27]

Artaud's view of the future was charged with pessimism for the future of man that could reject an inner sense, which he would characterize as the land of fire. Nor could Artaud find comfort in his own dark description of real theatre as a pestilence and terror for mankind. The cleansing power of his plague image absorbed and haunted him from his early teens through much of his life, cresting finally in his radio broadcast to the world in 1947, in which he blasted America for degrading the pre-Columbian Indian tribes.[28] Man must be reconstructed and scraped bare of "that animalcule that itches him mortally. Then you will have made him a body without organs, then you will have delivered him from all his automatic reactions and restored him to his true freedom."[29] Looking back at the achievements of both men, it might be said that, whereas Nietzsche thought deeply about theatre; Artaud, to his enormous sorrow, lived it.

How deep and central to life, to the nature of conscious life, was Nietzsche's vision of performance? As Marvin Carlson has paraphrased the German text, Nietzsche's theory of tragedy created a life-affirming response to the basic chaos of the universe. In Nietzsche's view, the distinctions between the Apollonian and Dionysian worlds: of dreams and of intoxication, were finally united in tragedy. "Ultimately," suggests Carlson, "the Apollonian drama is forced into a sphere when it begins to speak with Dionysian wisdom and even denies itself its Apollonian visibility."[30]

In stark contrast, Artaud's theatre offered no redemption from its apocalyptic vision of the spectator, plunged into the world of the plague, of torment, of bodily anguish. The balance in the Greek mind between the horror of the Dionysian world of existence and the counterbalancing Apollonian dream-world of Olympus that Nietzsche proclaimed, was a concept that Artaud was never able to embrace. The latter's vision of theatre has long been considered as inner-directed and self-absorbing. But Artaud's perceptions may not have been as narrow and restrictive as once thought.[31]

In the intervening decades since Artaud's death in 1948, strong evidence has been presented to the effect that the postmodernist experience led away from Dionysian intoxication and toward a theatre of multiple images, doubled, magnified, connected only by their linearity of intention. Now, with major research into the neurobiology of human perception of the last two decades, the Apollonian wisdom may be about to be displaced with discontinuous perturbations of deep Dionysian origins.

According to this conceptual line, the foundations of theatrical experience lie deep in the sub-cortical regions of the limbic system. Rational processes of the cerebral cortex exist to alert and to fine-tune the biological deterministic core. Any major problems arising are merely questions of interpretation by using information from PET scans and planar magnetic resonance imaging studies.

Computational mechanics of cellular processes, for example, have been introduced as a means of eliminating the nature of chaos in biological signaling programs. As of mid-1996, avenues of research were defined.[32] In the hard sciences, complexity theory had replaced chaotics as a fundamental building block of the universe. In the 1990s, speculation emerged that the nature of theatre might reveal the human brain as a complex adaptive system. However, it is one thing to announce a plan, another to implement the results.

Formal investigations, suggesting discontinuous adaptive signaling processes underlie all rational behavior, have barely been broached. The data from magnetic resonance imaging studies has produced a prolonged hiccup. These events are old news in a new guise: in the 1920s, Artaud, inhabiting that nightmare world where signaling processes were a living reality, experienced—and wrote about—his "Amygdalan adventures."

Claude Schumacher has written that Artaud hated theatre, but that theatre made him whole.[33] Theatre was life, lived (according to Artaud) with authenticity as the truth of one's being. For Artaud, role-playing was what happened in everyday life; the psychological drama, filled with an intellectual alchemy of feelings, was an artistic abomination. In Artaud's eyes, contemporary theatre of any value can only take place in the absence of "theatre." The artist must get beyond this superficial world of the critical faculties and engage Artaud's "real world" that existed at great depths in the mind.

Artaud's dichotomy between a super-imposed theatrical world and man's true theatre world posed a major dilemma. For Artaud, the true experiences of life in *his* real world can never be expressed in language, and yet language is the highest attainment of the species—a quandary that Artaud never resolved. Hence, intellectual passions (and the means of describing those passions) must be superseded. The question, "How can you justify the exclusion of spoken language in theatre?" haunted Artaud all his life. Postmodernist experimentation by such artists as Robert Wilson and Richard Foreman demonstrated the methodology. Only in recent years has medical science provided the rationale for a plausible answer.

Even in his last public pronouncements, the "vowel-speech" he delivered to astonished spectators in 1947, for example, Artaud's cacophony of sounds is more bizarre than revolutionary.[34] The speech gesture can be seen as an admission of failure and as a renunciation of earlier beliefs in the superficial nature of language. However, the cacophony can also be seen as a recovery of birdsong that dates back 65–230 million years. Hence, the reality of Nietzsche's pendulum swing between the animal and the Superman—above the abyss—is a quantum leap above Artaud's scuttled abyss. But if Artaud is given credit for astonishing insight into the dulcet male sex calls of mammalian reptiles (a long stretch), then Nietzsche's proclamation is merely a canyon updraft. In hindsight, it can be suggested that Artaud's apparent failure to communicate was, in reality, the more subtle success story. In a world "where nobody believes in god any more," Artaud's final quest was a remake of *Homo sapiens*. In the new anatomy, stripped bare of animality, man might cease to make war on himself and on others: "Then you will teach him to dance wrong side out as in the frenzy of dance halls and this wrong side out will be his real place."[35]

Ironically, since Artaud's death in 1948 his reputation has grown enormously, both as a prophet of the theatre, and as a seriously disturbed artist who nevertheless had something to contribute to late twentieth-century drama. The man who repudiated the intellect, whose visionary words (rather than whose theatrical productions!) were a major force in Modernism, would be astounded at this apparently perverse development. Indeed, the state of theatre in any given decade of the twentieth century might almost be judged in relation to its attitude to Artaud. The theoretical pendulum swings in favor of, or opposed to, Artaud's principles of dramatic art—either supporting his "irrational deep structures," or challenging his dismissal of "rational intellectual processes"—vary, much as the theoretical pendulum swing for and against Shakespeare as the "embodiment of drama" or as the "barbarian of theatrical forms."

Instrumental in the 1960s' development of Peter Brook's "Theatre of Cruelty" season at the Royal Shakespeare Company, Artaud appeared to be the embodiment of the theatre's living reality. As Bettina Knapp remarked, "The theatre of the Absurd, and the theatre of happenings, op, pop, and psychedelic art, electronic, serial and *musique concrète* were all designed to arouse, disturb, and evoke…visceral reactions capable of transforming or distorting man's conceptions of the world and himself."[36] A decade later, Susan Sontag decided that Artaud was a madman whose voice and presence was "unassimilatible," and that reading Artaud was "an ordeal."[37] In the early 1990s, Artaud was described as an artist who "loathed theatre."[38] How can we explain these readings of Artaud and an apparent breakdown of an essential paradigm: of Artaud as a prophet or as a madman of the theatre?

If we add in current social, political, gender and cultural issues to the Artaud debate, the issue is surprisingly clarified. In his authoritative master survey of dramatic theory, Marvin Carlson, for example, described other features of modernist and postmodernist theatre. Underlying its challenge to compete with other media for the attention of the general public, Carlson stated that 1960s theatre was torn between the need to be an engaged social phenomenon, and the need to be a politically indifferent esthetic artifact.[39] Bertolt Brecht represented the social phenomenon; the artifact was represented by Antonin Artaud. But Carlson's points of reference insist on an active engagement with society's external pressures. Bertolt Brecht's lehrstücks surely represented this perspective, in spades.

The problem is that Artaud is confounded and imprisoned by social, political, gender and cultural issues! The politics of the intellect demand that we consider artistic expressions in the theatre as products of social, political, gender and cultural issues. Artaud insisted that we abandon social, political, gender and cultural issues, and turn the psychic body inside out. What Artaud called for, the engaged esthetic artifact, the plague, the cruelty of "the bestial accident of unconscious human animality" that must be exorcised, is nonexistent.[40]

Since the middle of the nineteenth century, theatre has needed to express the latest scientific discoveries. The ultimate embodiment of this pressure to remain a vital and relevant "scientific" art form, beyond the scores of personal illustrations that could be listed, was Emile Zola's fascination with 18th century naturalists.

In the modern period, in the field of linguistics, for example, a driving force of contemporary dramatic theory concerns developed in two principal schools: the continental structuralism of followers of Ferdinand de Saussure and of Russian formalism, and the semiotic conceptions of the American, Charles S. Peirce. In recent decades, structural linguistics exploded in significance, with contributions from Eric Buysmans, Roland Barthes, Tadeusz Kowsan, Steen Jansen, Solomon Marcus, Mihai Dinu, Barron Brainer and Victoria Neufeldt, George Mounin, Andre Helbo, Umberto Eco, Marco de Marinis, Patrice Pavis, Anne Ubersfeld, Michael Kirby, Keir Elam, and Marvin Carlson.[41].

The conception that the dramatic text can be conceived as a linguistic process, and the stage itself as a logical, objective extension of the linguistic process, comes partially unglued when we realize that we are applying Newtonian analysis to Einsteinian problems, and Einsteinian solutions to Lorenzian problems of chaos and Weaver's complexity.[42] More particularly, in this text I try to distinguish and apply Warren Weaver's "ordered" complexity to problems of the real world.[43]

Semiotics consists of performance activities, made up of sign systems that are organized into particular meaning-bearing ensembles, with the aim of studying how the sign systems operate within a socio-cultural network, and how the object of interest (the actor as character for example) functions. The best exemplar of modern linguistic theory is Patrice Pavis, who, in Carlson's opinion, "sees the process of theatrical understanding as basically circular: the spectator receiving the complex messages of the stage begins to construct provisional codes, allowing the various

icons to be assigned stable signifieds. The signs thus constructed can change to signifiers of still other signs on the level of codes by the linkage of connotations...[that are] controlled by other primarily indexical signs, which point back to
the message...."[44] This linear description does not, however, even challenge the notion that a message can be seen as anything that can be objectively quantified.
Where does the message originate, and where does it return? What are the constitu-
encies involved in determining the rise and fall of information, and its forms, referential, mutable and immutable? What is involved in the spectator assigning codes,
and what is the difference between a provisional code and another, not provisional?
In the semiotic description of theatre communication, we are in danger of losing
the soul of theatre. We might as well describe the "recycling" of rainfall, or the "science" of weather forecasting.

Relativity to the situation, as conceived by each spectator, or by the direc-
tor/actor, as character, is born anew in every beat of production; as a consequence,
we create images that have no cohesion beyond the moment. As J. Dines Johansen
and Svend Eriksen noted: "the question is in fact how many of the elements and
events of the performances' spatio-temporal process can be interpreted as a coded
expression and which ones are contingent (i.e. incidental in relation to performance
considered as textual process)."[45] For these investigators, the narrow view of theatrical codes prevails, "how much of what is visible as also legible?" (that is, subject to
interpretation as meaning-bearing?).

The reality, as Johansen eventually notes, is that "the greater part of performance escapes analysis." Linguistic studies, semiotics in particular, have very little to
do with theatrical enterprises. There is also a real danger in cross-disciplinary studies, where extrapolations from one field of discourse point to certain functions that
are culture specifics only in a given historical context. Beneficial as the stretch of
new perspectives may be for a narrowly defined discipline, theatre in particular does
not respond well to any enterprise that focuses on intellectual properties of a subject
matter and fails to understand the true dimension of intellect in the theatre. Antonin Artaud probably understood this, intuitively, as did Nietzsche who, in *The
Birth of Tragedy*, lamented "the shrivelling of the full-blooded archaic theatre of
Athens by Socratic philosophy—by the introduction of characters who reason."[46]

The identification of space as a critical component of the cultural expression of
a civilization is a major contribution of semiotics. Marvin Carlson, noting that the
actor "always remains an uncanny, disturbing "other," inhabiting a world with its
own rules and which the audience, however physically close, cannot truly penetrate,"
gives up too soon. But Carlson's analysis of the anatomy of space is infinitely more
useful, where he announces that "a history of spatial signification in the theatre will
...be in large part a history of how different cultures have altered the location, size,
shape, and exact relationships of acting and audience spaces according to changing
ideas about the function of theatre and its relationship to other cultural systems."[47]

In the analysis of space, when we stay outside theatre function, semiotics has great use. But in describing the actor by going inside, by entering into the core of theatre as an expression of *Homo sapiens*, the best that semiotics can come up with is that this activity represents an "uncanny" and "disturbing" procedure. It's time to let the genie out of the semiotic box.

Chapter Five

◆❦◆

Consciousness and Craft

HALLUCINATIONS

The artist's consciousness in the act of creation has traditionally been off-limits. When I asked Sam Beckett how he put texts together, he looked perturbed. But when I told him of a man who, on reading W. B. Yeats' "Wild Swans at Coole," sank into a trance, Beckett was silent and absorbed.

This trance-like state is familiar to musicians as well as playwrights; In Aaron Copeland's words, the pattern emerges as "the inspired moment [which] may sometimes be described as a kind of hallucinatory state of mind; one half of the personality emotes and dictates while the other half listens and notates [music?]. The half that listens had better look the other way, and better simulate a half attention only, for the half that dictates is easily disgruntled and avenges itself for too close inspection by fading entirely away."[1] Copeland's description of the creative process is matched by other descriptions from many sources. Whether by hallucination, possession, or by relaxation, in the act of creation, something takes place that is beyond conscious, critical control. In summoning neural maps of ancient origin, in a creative fantasy of the kind Copeland described, the artist might well dictate the "call of Australopithecus" to "the other that half listens."

What is the goal of the actor? To become a madman? Surely not. Even Artaud might disagree. To live authentically? Consider Richard Dreyfuss's answer to the question, paraphrased: "Acting is childish. I create fantasies and live them. You must allow it to happen, to let go. Acting is also the most personal thing you will ever do, and you do it in public. You are completely exposed. But you must hold it in awe. Acting is awesome."[2] From a neurological base, Nobel Laureate Sir John Eccles has this to say about the emotional core of mammalian brains:

> Many behavioral patterns have their origin in 'feelings' penetrating the conscious brain. These feelings are deeply intertwined with demands of our internal organs and with the instinctive need for survival of self and species. Thus, the peripheral sympathetic motor system prepares the body for fight or flight while the parasympathetic section prevails over rest and recuperation.[3]

One can say the "awesome" work of the actor is to open the pathways to these systems. Eccles continues: "Stimulation of appropriate subcortical centers in the brain will precipitate complex behavioral patterns…. For example, stimulation of areas in the *limbic system, diencephalon,* or *brainstem* can provoke evasive or aggressive reactions in the absence of any external provocation."

Eccles speaks of artificial stimulation with drugs or probes, but stimulation can also come from actors activating 195 million-year-old selected pathways in the brain, "operating at a subcortical level." Mental gymnastics can also be duplicated by physical feats: in contemporary terms, Houdini's technique of dislocating his shoulder to escape a straightjacket is also the actor's art of "coordinating psychological and somatotropic actions" in what Dreyfuss might agree is equally an awesome performance.

ARTAUD AND CREATION

A living demonstration of neurobiological attributes of performance within the fractal boundaries of chaos theory, Artaud wrote as if his life depended on challenging inner demons. He spent his life describing what can be now described as "dissipative neural structures" and "deterministic chaos." Problems arose not with his writings, which are an almost helpless testimony, but with his everyday relations with the world, which both terrorized and stimulated his creative work.

The first-born in Marseilles, on September 4, 1896, to Euphrasie Nalpas and Antoine-Roi Artaud, a ship's captain, Antonin Artaud suffered a bout of Meningitis at age 5, which may well have underlain his life-long nervous disorders. Since this study is in part concerned with biological fractal boundaries, it can be noted at the outset that a recent postmortem study of schizophrenics by Eliot S. Gershon and Ronald D. Rieder typically shows *hippocampal* shrinkage, with a deficit in prefrontal blood flow, but no scarring. The suggestion by these researchers, is that "the abnormalities stem from a developmental disorder, perhaps a failure of the growth of neurons and the development of their connections, or from *a disturbance in the "pruning" of neurons that normally occurs between the ages of three and fifteen.*"[Emphasis added][4] Artaud's illness and subsequent behavior patterns fit within that neurological picture.

Within a decade of his birth, Artaud began suffering from periods of depression, and confinement in various institutions. All of these details are remarkably outlined by the annotated edition of *Artaud On Theatre,* with introduction and critical commentary by Claude Schumacher and felicitous translation by Schumacher and Brian Singleton.[5] Remarkably, this seminal figure in modern dramatic theory had a strong impact on directors and actors in the latter half of the twentieth century; and conversely, little influence on "Les enfants du paradis" during his lifetime.

A hint of what was to come can be seen in Artaud's verbal attacks on the Director of the Comédie Française. At the age of 29, with a range of experience in legitimate theatre already in hand, Artaud called for a new beginning. Railing against current practices, he proclaimed: "We aim beyond tragedy, the cornerstone of your poisonous old shed, and your Molière is a twat."[6] Earlier remarks to associates and to those whom he did not know reflect some of these ideas.[7] But here, in a brief passage, is Artaud incarnate. Intuitively, he had arrived at a statement of chaos theory of the mind. The rest of his life would be spent in trying to put this anti-theatre "mind" on-stage.

Of course this was impossible, despite the astonishing accuracy with which Artaud subliminally described what no dramatist has yet accomplished—the embodied and disembodied theatrical formulation of creation. There are fundamental reasons/laws within the discontinuous theory of chaos that argue against the possibility of ever reaching Artaud's goal. The chaos theory processes of anti-theatre as "the land of fire" are real; the anti-theatre as "solemnity" is to be sought: but you simply cannot pinpoint fractal boundaries for every customer—indeed, for any spectator.

In recent years, medical evidence in support of Artaud's instinctive cry against conventional theatre—from his perspective—has been impressive. Too much regularity, we have learned, destroys any biological system. The healthy heart does not march in synchronic rhythms. Intuitively, Artaud was right. Chaos, or plague, or disaster, in Artaud's colorful repertoire, is essential. Neurological conditions arise because, as doctors have discovered, there is too much order: "The complex rhythms of the nervous system had been replaced by a regimented beat or even drowned out altogether."[8]

Evidence cited by Cardiologist Ary Goldberger at Harvard Medical School shows that "the normal rhythm of the heart is surprisingly erratic.... The healthy heart dances, while the dying organ can merely march." At UCLA, kinesiologist Alan Garfinkel discovered that victims of stroke and of Parkinson's disease demonstrated regular bursts of electrical energy, while patients "with normal motor control had nerves that pulsed in a chaotic fashion." Similarly, Cindy Ehlers, a neuroscientist at the Scripps Clinic in La Jolla, California, notes that the normal person "will undergo erratic and relatively mild fluctuations in mood on an almost daily basis," whereas the depressed or schizophrenic personality displays a loss of some kind of control mechanism, "so that over time their daily behavior starts to look extremely periodic or rhythmic."[9]

It would seem that a rhythmic biological regularity leads to madness, whereas a little bit of chaos is extremely beneficial to the biology and mental stability of the species. The fact that these experimental results are exactly the opposite of what many would consider to be the logical intuitive response is further evidence in favor of complexity theory as a foundation of physics and of the natural and biological sciences. Once again, mammalian biology tells us that the rhythms of life are underscored by *movement* within those beats.

From another perspective, Bettina Knapp detects a curative agent in work in the art of Artaud. She sees Artaud's theatre as "a means whereby the individual could come to the theatre to be dissected, split and cut open first, and then healed."[10] The Frenchman's ideas concerning the dramatic arts "were born from his sickness" and then "thrust upon the spectator." At some succeeding moment, in a gigantic catharsis, the spectator would come to "a recognition of Artaud's intention to force the spectator to recognize the nature of his projections and anxieties." Having gained this greater perception and self-understanding, "the fragments of the spectator's personality. which had been projected onto the stage," could now "return to their source, the spectator's being—nourished and renewed by the added understanding."[11]

The only difficulty with this interpretation, which might well be true, is the demands placed upon the spectator to engage the stage demons. Doubtless Artaud would demand this; doubtless, few spectators would take the trouble to oblige. (There is a parallel to Artaud's work in Robert Wilson's demands for active spectator involvement; but whereas Artaud's verbal assault is gut wrenching, Wilson's approach through imagery is cool and translucent.)

Artaud clearly did not know the dimensions of his own creativity. Embroiled in controversy from the beginning of his career until his confinement in Sotteville-les-Rouen in 1937, thence to Rodez in 1943, where he remained until his release to friends in Paris in May, 1946, Artaud's work spilled over the banks of all probity. Claude Schumacher reports on a 1971 statement by one of Artaud's physicians in Rodez, Dr Gaston Ferdière, who wrote:

> For I had certainly not "cured" Artaud and he was really incurable with the present resources of psychic therapy. And I don't understand what the words "cure" and "mental health" mean in relation to such an exceptional man.[12]

Certain other critics—Susan Sontag in point of fact—have not been solicitous of artistic excesses. Pointed questions may be raised on the legitimacy of Artaud's sensibility when he was in his "right" mind. But that should not justify her dismissal of Artaud outright: "It is not a question," Sontag stated, "of giving one's special assent to Artaud—this would be shallow—or even of neutrally 'understanding' him and his relevance." And she proceeds to mount her own polemic:

> What is there to assent to? How could anyone assent to Artaud's ideas unless one was already in the demonic state of siege that he was in?.... Not only is Artaud's position not tenable, it is not a "position" at all.[13]

As there is no curative for Artaud's particular madness, there is no curative for the literary lioness that denounces the poet for living life in the interstices of consciousness.

Curiously enough, it was Artaud's own language that interfered so devastatingly with his message to abolish language in the first place. The text of *The Theatre and Its Double*, and Artaud's letter "To the Director of the Comédie Française," (Paris, 21 February 1925) might serve as his manifesto to the century: "The theatre is the

Land of Fire, the lagoons of Heaven, the battle of Dreams. The theatre is Solemnity…. You leave your droppings at the foot of Solemnity like the Arab at the foot of the Pyramids. Make way for the theatre, gentlemen, make way for the universal theatre, which is content with the unlimited field of the mind."[14]

In his addresses to the world at this stage of his life, Artaud still possessed a kind of child-like obsession with sexual innocence. The public pronouncements of personal pain and anguish, the need to excoriate sexual fantasies, the muddled invective that so detracted from any artistic insights, followed in his broadsides in the next two decades. Very much like Oedipus's recommendation to mankind from Sophocles, Artaud's least favorite Greek poet, wisdom through suffering did come to Artaud, but not in a tolerable form.[15]

INTERSTICES IN THE CHASM: THE MEDICAL EVIDENCE

Samuel Beckett describes reality in the interstices of language. The best proof may come from medical literature. The case involves the brain of Karen Anne Quinlan. We believed for centuries that cerebral processes govern our activities and keep us sane and rational. However, in 1994 medical science suggested that part of the Proto-reptilian brain of man, the old brain, is critical for cognition and awareness, and is absolutely essential for arousal states. This assertion threw out comfortable beliefs of generations of modern researchers. Theatre practitioners readily assume that cognitive acts are made by our *cerebral cortex*, but medical evidence suggests otherwise.

Karen Ann Quinlan survived for 10 years in limbo. With a normal functioning *cerebral cortex*, she should have had a wakeful, happy life. Her brain scan was normal. Electroencephalograms showed cortical activity when the patient was awake. But, having ingested prescription drugs and alcohol, she suffered cardiac arrest and slipped into a "persistent vegetative state."[16] She slept and awoke each day, but was unable to exhibit any signs of thought, emotion or awareness of her self and surroundings—certainly key elements in the typical interpretation of a role for the stage. The damage was deep in her mid brain, and that damage prevented her from having a full life.

In brief, Ms. Quinlan presented to the world a state of chaos without order, a latent consciousness: a chasm of being, a deterministic biology, an intentional matrix which no-one could decipher, or cross or enter and examine in the dynamic sense of an active, thinking, human being. Clearly, there is more than chaos theory at work here. Not until dynamic principles of complex adaptive behaviors are understood, and the agencies that function at the deepest biological level are comprehended, can the nature of Quinlan's consciousness be addressed. Is the Quinlan conundrum a subset of Artaud's theatrical consciousness? Can answers be presented, perhaps as a computer simulation of a multi-million run of several billion neural maps of the proto-reptilian brain?

If the *cerebral cortex* and, in particular, the *frontal lobes* are not the executive function of consciousness, what roles do the old brain play in the alert being? Is the *thalamus*, or the *amygdala* "the other" in consciousness? The I in "not I" that Beckett speak of in his play of the same name, or the unnamable in his novel, *The Unnamable*, may be the irreducible zero sum of consciousness that we seek without knowing and find without searching for.

BOUNDARIES OF THE TRIUNE BRAIN

The suggestion that the human brain has a hierarchical structure, related to evolutionary behaviors of man and his predecessors, remained an unproven hypothesis until a young woman in a permanent vegetative state (as a result of a drug overdose and excessive alcohol), was autopsied. In the case of Karen Ann Quinlan, the medical team, having sliced her paraffin-preserved brain very thinly for microscopic examination, used computers to produce three-dimensional pictures of the damaged core. An unexpected test of Paul MacLean's thesis of the "triune brain" was the result.

Consisting of the entire *thalamus, hypothalamus, basal ganglia, hippocampus, amygdala, basal forebrain, temporal lobe, posterior frontal cortex,* and *parietal cortex,* the sample core revealed extensive damage in the proto-reptilian brain. Previous research had shown that persistent vegetative states such as the one Karen Quinlan was in occurred with severe *cerebral cortex* damage. But Quinlan's cortex was intact, while her *thalamus* (old brain) was severely damaged. The only temporary conclusion taken from this evidence was that the *cerebral cortex* is not the root of cognitive processing, but only of fine tuning of cognitive processes that take place in an area of the brain that is up to 200,000,000 years old. That is not to say that the *thalamus* might also be a conduit of other old brain functions, in addition to its service as a conduit of neural activities to the *cerebral cortex*. The reverse might be true. For example, the *amygdala* is triggered by signals from the *cerebral cortex* to stimulate the *thalamus* into action. We have only begun to unlock those mysteries of the deep origins of cognition and awareness.

The presumption could now be made that we think in our proto-reptilian brains; We process thought and turn it into language and comprehension in the *angular gyrus* of the *cerebral cortex*, a part of the neo-mammalian brain, but we do so under the influence and control of the *frontal lobes*. What the researchers did not state is the extent of damage to the neural connections from the *frontal lobes* to the *thalamus* region and back to the *temporal left lobe*.

Before language, before thought is put into words, there is cognition and awareness. Arousal is of another order of thinking. The *secondary association area* bounding the *angular gyrus* in the left *temporal lobe* is a unique development of man's triune brain. I am able to express what my brain interprets from both the internal and external world. In a sense, this might be considered the beginnings of a new brain. And if neurobiological history is any indication, laterality is a key ingredient in evolution. A similar *angular gyrus* will arise in the right hemisphere in another million years, perhaps trained in abstract visual thought processes.

The question is open-ended as to what use the right hemisphere will find that will perhaps make Einstein's Special Theory of Relativity mere child's play, or computer simulations of complex systems as mundane as tabulations of an adding machine. Certainly the Quinlan research and the more recent research into the functions of the *prefrontal lobes*, by Elkhonon Goldberg and others, seems to vindicate Maclean's proto-reptilian triune brain dynamic.[17]

This discovery has profound implications for theatre. We can now proceed to open up communication at the boundaries of language. Using descriptive language from chaos and complexity theory and integrating Edelman's insights into functional networks of the human brain, a short summary might be presumed as follows:

> Neural maps constructed by dissipative systems, with fractal boundaries of neuronal networks in the discontinuous *prefrontal cortex* of the brain, with reentry connections to the *thalamus*, are the core of thinking.

The key element in all of these equations is a discontinuous *cortex*. The infinite ability to adapt to changing environments allowed a singular creature to survive and adapt into other species for 195 million years. The entire process took seven and a half billion years to sort out, as I have suggested, plus a number of galactic strikes from outer space, just to keep the species on their evolutionary toes. Once there, nothing could stop this creature (*Hadrocodium wei*) from thriving and evolving, where other species came and went. Artists in one guise or another have always had intimations of this origin. Picasso—in describing the roots of artistic creation as *Cosa Mentale*—the thing that happens in the mind—or Beckett, in speaking about interstices in language, explore the range of consciousness, and bring us to the boundaries of the ancient and the irrational.

What has always been true of great art is now also true of medical neuroscience. The envelope of theatrical consciousness has opened a little more. The role of the *thalamus*, which has heretofore been without a voice, an irrational shadow in the mind, a relatively insignificant part of the old *mid-brain* of the species, is crucial to all cognitive processing. The *thalamus* is the core of the theatre's double. The *thalamus* cannot speak for itself, it cannot identify or be known for itself, but it is the voice through which an ancient species quite unlike any of us speaks to us.

When I hear an actor speak of the pain of true emotion, of finding an authentic voice, of the anguish and private struggles to speak with a public voice of a character; when I hear Jessica Lange, for example, "bare all" to the Actor's Studio, in a revelatory confession of her technique, of her need for technique, because she cannot get there herself, I know I am listening to someone who has been to the well and found through her pain the freedom to express the nature of creation that artists can bring to us. That is what art has to give to us, an ancient movement filtered through current consciousness. That is what we go to the theatre to understand. Research now shows that the signal event that transformed creatures like ourselves into thinking beings was the emergence of the *cortex*. As Goldberg notes,

> Throughout evolution, the emphasis has shifted from the brain invested with rigid, fixed functions (the *thalamus*) to a brain capable of flexible adaptation (the *cortex*).[18]

And so the possibilities of an infinitely adaptable creature emerged about 65-plus million years ago through the intercession of successive meteoric strikes at the earth. When we create theatre of great interest to others, we reach back to the fixed functions of an evolutionary period hundreds of millions of years old, to a paradigm shift from interactive brain function to modular functions, and then back again. The process is called ordered complexity.

Part Two

❧❦❧

Ancient Wonders

Primitive languages consist of very long words, full of difficult sounds and sung rather than spoken... The early words must have been to present ones what the *plesiosaurus* and *gigantosaurus* are to present-day reptiles.

—O. Jespersen, *Language*

Chapter Six

❦

Complexity and Discontinuity
of Species

I think I have found out (here's presumption!) the simple way by which species become ex-
quisitely adapted to various ends.

Charles Darwin, Letter to J. D. Hooker, 11 January, 1844[1]

A THALAMIC-RIDDEN SPECIES

Charles Darwin waited decades before publishing his 1859 *Origin of Species*. The product of his expedition aboard the HMS Beagle to the Galapagos Islands, he originally set out to prove the literal truth of the creation of species in the Christian Old Testament. Darwin's conclusions were exactly the opposite. Creation of immutable species did not occur in 4004 B.C. at precisely 9 A.M. (Was this Greenwich Mean Time?)

Even in finding the courage to publish his findings, in the face of his strong Christian faith and social standing in his Down, Farnborough, Kent community, Darwin could cry out in despair against a God who could leave his creatures, across all time, to evolutionary fate. But even that cry of pain could not match the real despair of the human species, priding itself on civilized behaviors, and elevated cultural standards, subservient to the rule of law, yet still a creature of reptilian appetites.

I cannot conclude this section without noting that Darwin's presumption of species' succession and adaptation could be based on nineteenth century capitalism practices, rather than an anomalous biological accident that gives certain species a temporary advantage. "Survival of the fittest" might well be "serendipitous evolution." Complexity theory (and religious fervor) would argue for the latter.

A question for the ages: if there is so much communication and knowledge of humans between and about the human species through evolutionary development, why is there so much war and civil discord, mayhem and murder, rape and pillage of millions of innocent victims, across the history of the species? Those who would claim that something went wrong with the development of human species need to be reminded that *Homo sapiens* have evolved quite nicely. You cannot obliterate the post-dinosaur evolutionary track of 65 million years, or of the post-Cambrian Epoch of 200 million years, or even of a plan that began 7 1/2 billion years ago in the emergence of our second-generation solar system. Nevertheless, the capacity to create theatre is also the capacity to do good and evil.

To comprehend that pre-history we have theatre and theatre artists who are willing to undertake explorations of old rhythms and ancient energy configurations that predate the species. In learning about these forces in contemporary guises, we learn about the full nature of ourselves. Hence theatre is the most civilizing force in society, for it permits a deep investigation of neural forces in a discontinuous forum, that of spectators allied to each other in a group setting, but able to participate in a uniquely individual manner.

While Darwin examined data on plants, sea animals and inhabitant of the Galapagos Islands for his epoch breaking text on evolution, he was presciently mindful of the real basis of study: the challenge to biblical texts he left for the next generation of scholars to ponder. When one looks at the span of theatre—from *Gilgamesh*, circa 2300 B.C. to dithyramb contests and Peisistratos in 5th century Greece; from the Pharoah plays of Egypt, circa 3200 years before the present, to *Horus*, 12th century B.C.—down to the present day, a scant 5200 years—the line is discontinuous, reflective, symbolic of man's higher cerebral function, and all contained in language.

This record is not the beginning of theatre but the beginning of language as recorded history. So, the question might arise, where is the beginning of language, and possibly, the beginning of recorded consciousness? There are no endocasts, as noted, of speech. But there are patterns of speech, or voiceprints, which can serve as the missing links of consciousness in the absence of Paleolithic evidence.

Through language we can trace the emergence of *Homo sapiens* out of Africa and through the entire world. We can follow the emergence of new language and the discarding of the old, and we can trace, though language the initial arrival and evolution of consciousness as a species-unique phenomenon. It may be assumed that the human race produced a range of vocal sounds with a variety of intonations long before language began.[2] The types of "first words" were products most importantly of the discontinuous cortex. Even though birds and humans are blessed with fluent complex sounds, described by Aitchison as "double-barreled and double-layered" systems involving tunes and dialects, the human variety is centered in the *secondary association area* of the *angular gyrus* in the left *temporal lobe*. Other similarities include a juvenile sub-language in birds and man, and a disposition to acquire the system more fully in the juvenile stage of life.

Important differences include the interesting fact that only male birds sing, while female birds are songless unless injected with testosterone. The origins of song in mating calls of the male has spread in humans to the female of the species much as the birds' annual mating instinct survives in humans as a perpetual receptivity cycle. Of the various types of first words, Aitchison notes four distinct categories: "heave" types, "Mama" sounds, animal imitations, and *ontogeny adaptations*, where there is a distinct physical development that is evolutionarily advantageous. Aitchison notes the lowering of the larynx as advantageous to the freeing of vocal sounds. Similarly, there is the development of a "naming insight" as a mental change, and therefore the development of a symbol system that eventually emerges as language.

Darwin understood the implications and very precisely formulated the evolutionary track that man was on, and believed "that only a few species of a genus ever undergo change."[3] He continues:

> It is a truly wonderful fact—the wonder of which we are apt to overlook from familiarity—that all animals and all plants throughout all time and space should be related to each other in groups, subordinate to groups, in the manner which we everywhere behold—namely, varieties of the same species most closely related, species of the same genus less closely and unequally related, forming sections and sub-genera, species of distinct genera much less closely related, and genera related in different degrees, forming sub-families, families, orders, sub-classes and classes...can be explained through inheritance and the complex action of natural selection, entailing extinction and divergence of character.[4]

This is not Artaud disparaging the race, but a scientist expressing wonder. Darwin was surprised at the biological advantage that would accrue to a species that we now know has a discontinuous cerebral cortex. In the act of reflection, the tiny forest species predating man gained a slight advantage, a "theatrical advantage." Where other similar species were extinguished, Hadrocodium wui's bio-evolutionary offspring adapted to galactic missiles, climactic changes, species constraints and ecological threats to become the dominant species on earth.

As noted earlier, to act, in the case of theatre, is to respond to movement. The basis of all theatre, according to Aristotle, is *mimesis*: "imitation of an action." In the theatre, man portrays the history of his or her evolutionary past dating back to a reptilian core. "Movement" of a neural impulses across a synapse will alter any perception, and is just as real as any external mimesis with its own ordered complexity. Emotions dredged up by actors have a real basis in pain or joy, for they present to a spectator the resolute determination to comprehend—and to triumph over—irrational man across the entire geological record.

At the beginning of a new century the origins debate rages on. Evelyn Keller of MIT disputes physicist Freeman Dyson's suggestion that life originated as a symbiosis between a self maintaining metabolic system involving proteins and a population of inaccurate replicating molecules, probably nucleic acids. In contrast, John Maynard Smith of Sussex University, UK, believes that the first living things—"that is, the first entities with heredity and so able to evolve—were molecules, perhaps RNA (a molecule resembling DNA but single-stranded), which acted both as inaccurate replicators and as primitive enzymes...that evolved gradually into a DNA-protean system, with a genetic code."[5] How amazed would Darwin be to see at first hand how evolution has progressed into what is now an "ordered complexity" theory origin debate.

LANGUAGE AND EVOLUTION

Consciousness has no endocasts. There are no bone fragments or skeletons to unearth, no monuments or tombs to revere "the first user." But we do have speech patterns and evolutionary changes in language use as living endocasts of consciousness in action. For the moment, speech is the best evidence of modern consciousness as we know it today.

In the beginning was the noun, followed by a single word, usually a verb. From that came syntax and the beginning of confusion. Aitchison explains, "First, language made use of the world, as filtered through human experience. Above all, it made use of the human body and its location in space to move outwards to the surrounding environment, and inwards to inner ideas."[6] Presumably the human race possessed sounds with a wide variety of intonations, according to Aitchison, long before it acquired language as speech. Once a syntax rule was acquired, noun plus verb or verb plus a noun, the generative principle of communicating an inner sense of self to the surrounding community developed into a need to clearly define that inner intention. As a result, language acquired additional parts of speech, such as prepositions, word-endings, and postpositions to take care of multiple meanings.

What is particularly valuable in human speech is the discontinuous and generative power of sound-images and sounds interactions of finite resources to produce an infinite variety of sentences. No other species has this generative capacity for language, and for word-generation within language, to form other words and other images. Not since bird species, those ornithological links to the dinosaurs, went their separate evolutionary ways prior to 65 myBP, has the human species' language production had a common origin.

Of equal importance is the rule-bearing character of languages in every region of the world. As the species spread across the globe, Aitchison believes that mass communication through language became possible soon after or at the same time. The birth of consciousness marks the dispersal of populations, in the form of small bands or families, across the continents. The estimates vary from 75,000 years to 120,000 years ago, with much of Europe occupied by 45,000 years before the present, the Middle East by 50,000 years before the present and to Asia and Australia by the same date. The Americas were the last continents to be inhabited, across the Bering Strait by 30,000 years before the present, and thence down into Central and South America.

The notion of a sudden surge of neural activity in man, a "punctuated equilibrium" in consciousness, unleashed by the receding ice age, that literally exploded out of a very small area of the African continent, cannot be ignored. More likely the consciousness quotient originated in a small band of wayfarers once they had left Africa, became isolated in the ice age, and adapted to the harsh environment. Through an evolutionary quirk that enlarged the association areas of the angular gyrus of the left temporal lobe, sound-images were available as a basis for communication. Quite suddenly, in geological time, this anomaly became a supreme advantage for hunters to track their quarry, for mothers to nurture their offspring, for communication with other members of the tribe to take place. In place of minimal vocal-sounds communication, language followed, and language-communities dispersed throughout the landmasses.

How can these predictions be developed? Not though fossil remains but through "language artifacts." Johanna Nichols at U.C. Berkeley calculates a stock or major language grouping lasts 5,000 years and fewer than two daughter languages are likely to survive from it.[7] Aitchison explains how the historical equation is arrived at:

> Approximately 140 Amerind (American-Indian) stocks now exist, so 5,000 years ago there were half that number, or 70 stocks, 10,000 years ago only 35, and so on. It would take 35,000 years to reach a figure of under 2 stocks. The New World must therefore have been settled at least 20,000 and maybe even 40,000 years ago.[8]

Aitchison has developed a methodology, known as "comparative historical linguistics," out of its nineteenth century roots. By examining "daughter" words of a known parent language, a history may be deduced. There is also a system called "population typology" which examines dissimilar linguistic features: the inclusive/exclusive opposition of "we" as "you and I," versus "we" as "I and some other people." This opposition occurs in 10% of European and Caucasian languages, in 56% of South and South East Asian languages, but in 89% of Australian languages.[9]

Using this methodology, Nichols has been able to estimate the outward expansion from Africa of humans. The first stage begins in Africa to the East of the Great Rift Valley; the second stage sees human beings colonize Europe, inner Asia, Australia and eventually America, circa 60,000 to 30,000 years before the present. The third stage involves the rise of complex societies, which "leads to a reduction of linguistic diversity, and the spread of a few lineages across the known world."[10] Of great interest is the Eastward increase in diversity, since this indicates a very old pattern relatively unaffected by complex Western influences. Hence our purest sources of original language and of the consciousness that nurtures language, come from the Pacific and New World, and still reflect the very early population expansion.

In no cases do the language endocasts find origins earlier than 120,000 years before the present!

CAMBRIAN ATTRIBUTES OF CONSCIOUSNESS

More dramatic evidence of an earlier effect of punctuated equilibrium occurred with the demise of the dinosaurs, probably as a result of a collision with a giant meteor in the Yucatan Peninsula in the Gulf of Mexico. Lacking sunlight and plant life, the dinosaurs vanished within a period of months, if not years. If they had not, the intelligent forms of life on earth might have been *two-legged, big-brained, mammalian reptiles!!* In their absence, a niche in the evolutionary shelf was opened. As with the creation of the universe and material within the universe from the remnants of the "war between the quarks" that led to successive genesis of super novas, and the formation of a third generation star (and our sun, and bits of the sun's debris to form the planets), the almost inconceivably thin line of *Australopithecus africanus* barely survived its own catastrophic period at 2.5 myBP to carry the genes of modern man to the split of *A. robustus and H. habilis.*

Looking back, it is clear that *Homo sapiens* faced incredible odds to become a species when every branch in its historic genetic line, with the exception of *H. erectus,* met with oblivion. The incessant danger of living through a succession of primeval epochs, and the wit to survive and even to flourish, is a part of the heritage that underpins the neuro-psycho-physiological basis of man's culture; theatre being its most basic expression, and possibly its highest attribute.

Descendent of mammal-like reptiles that trace their ancestry some three hundred million years ago, whose primogeniture appears to be a kind of fish called *Crossopterygian* (350,000,000 years before the present), the mammals entered the age of dinosaurs some two hundred million years ago, and quickly lost the battle—almost. Carnivorous dinosaurs decimated mammalian life. Survivors shrunk in size, until the typical mammal was no larger than a mouse, creeping across the forest floor, taking shelter during the day, foraging at night. However, this creature had a remarkable advantage—an outsized brain—even for the period, which predicated a capacity for innovation and adaptability to the environment in order to survive.

More remarkable, this tiny mammal—*Hydrocodium wui*—developed a "smell brain," by which it navigated for food and water.[11] Vision for a night animal was no longer of central importance. A sense of smell and of hearing dictated whether the animal would survive in its nocturnal habitat. (Note that even today, a sense of smell in man is uniquely a *limbic system* response, and not a *cerebral cortex* function.) Gradually a coating of grey matter grew over and covered the "smell brain," where the animal conceived of its plans to attack, to foray and to retreat from danger as necessary. The growing grey matter covering the area, the beginnings of what would become the cerebral cortex in man, now accentuated any plan of action deriving from the smell brain.

Perhaps just as important, the fossil bone structures of the ear indicate these tiny mammals—our mammalian forebears—had better hearing than their reptile ancestors did. The grey matter, or *neopallium*, was used to coordinate smells with the other senses to arrive at a course of action. As Marcel Proust could testify, in *A la Recherché du Temps Perdu*, the sense of smell goes directly to the brain and allowed the forager to out-scent his prey or his tormentor, just as it allowed Proust, (or the contemporary artist) to accurately re-imagine entire scenes from the smell of a magdalene dipped in an infusion of tea.

While the brain of mammals continued to grow, eventually dividing into two bumps on either side of the smell brain, to later become the two cerebral hemispheres of the human brain, reptiles began to grow to enormous size. Certain fossil records indicate that dinosaurs had markedly diminished intellectual resources. From a biological perspective, and in the absence of serious threats to their well-being, intellectual prowess was superfluous. Recent investigations suggest that these gigantic creatures, beginning in the Triassic period (248 myBP), through the Jurassic (213–144 myBP), and through the Cretacious period (144–66 myBP), exhibited the same brain size and developmental patterns as the tiny mammals. In other words, the reptile bodies had far outstripped their reptile brains.

Some dinosaurs developed remarkable talents for nesting, herding and hunting in groups. One of the last—and the smallest tyrannosaur, *Nanotyranus* (sixty-five million yBP), had a brain twice the expected size, a large vision center that would resolve minute details, and a sensory gland that was probably as efficient as today's modern predators. However, an older counter-argument also suggests that mammalian brains were typically five times larger than *Tyranosaurus,* and proportionally twenty times larger than herbivorous dinosaurs. A huge body and a tiny brain meant that dinosaurs used the entirety of their thinking process for locomotion—to find food and to flee from predators. In contrast, by one hundred million years ago, the little mammals were successfully adapted species with large brains, an improved body design, warm-blooded metabolism (sufficient to provide energy to their demanding brains and to survive in a cooling climate), and with an instinct for parental care. The recipe for discontinuous neuronal connections that could result in intelligent thought processes had begun. Theatrical consciousness, the ability to reflect and draw abstract conclusions, and to act on those judgements, might take another hundred million years, several attacks on earth by interstellar space rocks and a few ice ages to mature. But the die was cast.

In the interim, these mammals were still not a match for surviving reptiles and huge dinosaurs. By eighty million years ago, the continents began to drift apart, weather became more temperate—cooler and drier—than in the previous one hundred million years, and the question of adaptability of species became a central issue. Mammals adapted, dinosaurs could not adapt. Alternatively, some dinosaurs' adaptations to climatic changes included a reflexive metabolism: during early development, warm-blooded dinosaurs grew quickly. On reaching maturity, their metabolism slowed down, and they cooled down metabolically.

The end of the dinosaurs came not with an inability to adapt to change—adaptations were constant through the Mesozoic Period through minor meteoric strikes that wiped out the dinosaurs' major competitors—but through a catastrophic asteroid smashing into the Earth. This late Cretaceous equivalent to our "nuclear winter" doomed plants and the animals that ate those plants. By sixty-five million years ago, the age of the dinosaurs ended. Their small flying cousins—ancestors to modern birds—and mammals quickly filled the evolutionary niche, and rapidly developed within their new ecological place a greatly modified "*neoreptilian*" brain.

That this evolutionary tale is of consummate interest to artists, detailing the rise and fall of species and their biological genetic inheritance to our time, is abundantly clear, from Picasso's African Period, to Maya Angelou's Presidential Inaugural Poem, *On the Pulse of the Morning*:

> But today, the Rock cries out to us, clearly, forcefully,
> Come, you may stand upon my
> back and face your distant destiny,[12]

In Samuel Beckett's *The Lost Ones*, "the ladders" of biological evolution of vertebrates are described on a universal scale and the genetic evolution of succeeding generations of man and woman are formatted on an intimate DNA scale:

> And yet it takes courage to climb. For half the rungs are missing and this without regard to harmony…. The purpose of the ladders is to convey the searchers to the niches…. They mount to the level of their choice and there stay and settle standing as a rule with their faces to the wall. [13]

Speculative stereotypical behaviors in the proto-reptilian brain include the establishment of home territories, the finding of food and shelter, as well as breeding rights. In the *paleo-mammalian brain*, nature's first steps were taken to provide a sense of self-awareness, and especially an awareness of the internal conditions of the body. This "visceral brain" setting for the *hippocampus*, the *basal ganglia*, and the *cerebellum* developed a capacity for significant memories of the inner and outer world. Internal information came from the *septal areas* of the brain; external information arrived from sensory systems that projected into nearby transitional cortical areas. In the late development of the neo-mammalian brain, responsibilities for cold (non-emotional) and fine grain analyses of external environments, developed in the *cerebral cortex*, an interlaced six-layered sheet of about ten billion billion (a thousand trillion) connections.[14]

Edelman notes specific functions of the *paleo-mammalian brain*: the *cerebellum* surrounds the *brain stem* and receives two main kinds of inputs from the *cerebral cortex* and the spinal cord, which allows it to play a role in the timing and smoothing of successions of movements; the *basal ganglia* functions in the long range execution of sequential motor events that connect to the *cerebral cortex*: in particular, sequential eye movements and body movements related to behavioral planning and emotional displays which are linked to frontal portions of the cortex. Finally, the *hippocampus* has important connections to hedonic centers of the *mid-brain* and is the site of regulatory mechanism for changing short-term memory to long term.

All this evolution had to take place before consciousness could even be speculated on in any creature. It may well be that consciousness takes about ten billion years to evolve in a planetary mass. With a few serendipitous asteroid collisions, planet earth did it in seven and a half billion! In this triune brain model there are a number of checks and balances: the *paleo-mammalian brain* frees the *proto-reptilian core brain* from certain starting mechanisms, thereby integrating the creature's over-all response. Early computer simulations of our brain faced the same problems. Every time a new "thought" was introduced, the computer had to start from scratch. With the advent of parallel processing, that particular problem disappeared.

In the biological brain, the *hippocampus* has the capacity to suppress activity in the proto-reptilian brain when the unexpected happens. Conversely, the *amygdala* serves to heighten conditions of awareness and activation of the *hypothalamic* system when external conditions warrant extraordinary response mechanisms. And finally, *prefrontal lobes* of the *neo-cortex* operate as a discrete system by activating *secondary associations* of the *inferior parietal lobe* to interpret external stimuli through "consciousness."

In the lower vertebrates, the nervous system acts to regulate bodily functions and behaviors. In the precursors to hominids, the evolution of the cortex allowed for perceptual categorization and for the ordering of these perceptions into sets of psychological functions. This latter advance led in the early *Homo* species to the ability to form concepts, the certain precursor to speech.[15]

Seen from another triune perspective, Robert Jastrow summarized his own interpretations by stating that the *proto-reptilian brain* was itself divided into three compartments, a frontal compartment for smell, a middle compartment for vision, and a rear compartment for balance and coordination.[16] These compartments grew out of the brain stem, and much of the repertoire of human responses to the world goes through these deeply buried regions that directed our ancestors hundreds of millions of years ago.

In Jastrow's view, receptors for vision and smell were coordinated in a region between the smell brain and the vision brain. Called the *diencephalon*, this area became in advanced mammals the site of the *thalamus* and *hypothalamus*. The basic limbic system instincts for survival, sexual desire, searches for food, anxiety and aggressive responses, are wired in his area. As mammals moved into the daylight, a new improved system of vision added even more neural circuits to the cerebral hemispheres. The capacity for thought processes expanded exponentially and eventually developed into a truly conscious, thinking being, capable of creating an external sense of self in visual and plastic arts, and in articulating that self in language and writing.

No originatory dates are available, but speculation can suggest *proto-reptilian* genesis as three hundred million yBP, *paleo-mammalian* as one hundred and ninety-five million yBP, and *neo-mammalian* as two and a half million yBP. What we are speaking of is a complex system of checks and balances that operate at the biological level, a kind of "native intelligence" that develops into "consciousness" in *Homo sapiens* as recently as one hundred and twenty thousand years ago.

Linked inextricably to the creativity of man, perhaps defining the species most acutely from all other creatures, is the extraordinary *plasticity* of *Homo sapiens* to environmental changes. We are creatures of the ice ages that have dominated the earth for 2.5 million years. In the geological record, mass extinctions due to asteroid collisions have occurred five times in the past five hundred million years. Two hundred and fifty million yBP, another collision on the boundary between the Permian and the Triassic Period destroyed 80 percent of all creatures on earth and began the ascension of the dinosaurs. Dinosaurs survived the next galactic assault, circa two hundred million yBP, but not the next impact, sixty-five million yBP, which began the Cretaceous Period.

With minor extinctions every twenty-eight million years—the last coming some ten to eleven million yBP, (not coincidentally, the exact time of the split between pongid/hominid "apes" and hominid "man")—our solar system is beginning to appear to be a regular complexity system of its own. Little is known of the earliest giant space rock impacts, 420 and 350 million years ago, respectively. We'll know after another million impacts, if our descendents survive and replenish.

SECONDARY ASSOCIATIONS IN THEATRE

From a DNA perspective, *Homo sapiens* evolved out of *Homo erectus*. Quite possibly as a result of "sprung evolution"—Jay Gould's 'punctuated equilibrium'—a particularly advantageous mutation during the current ice age was replicated (and reinforced) in a small, focused community.

The process of creating a mutation is intriguing. The genetic code is composed of strands of DNA. Each strand of DNA contains a long string of four smaller molecules called *nucleotide bases:* guanine, cytosine, adenine, and thymene.[17] Each base contains a complementary strand wrapped around it. Complementary bases pair with each other: G with C; A with T. Within each strand there are strong chemical bonds; across each strand there are weak chemical forces. Across strands, the bonds come apart readily, with heat, etc.; within each strand the bonds are strong. By using any strand as a template a second strand can be built from a single base by special protein enzymes, with the order in the new strand determined by complementary pairing. In this fashion, a sequence of three bases (G, C, A or T) represents a *code word,* telling the cell to incorporate an amino acid into a long string of such amino acids, called polypeptides. This newly created polypeptide folds up to form a protein.

Now it gets complicated to follow the pairings. If each *code word* consists of three *nucleotides*, then 4P3 = sixty-four code words can be constituted from four kinds of nucleotides. A piece of DNA of the right length and base sequence to specify a protein is known as a gene. When a cell divides, it copies the DNA from one of the RNA strands to provide new DNA for its daughter cells. However, in the case of sprung evolution, if there is a mistake or a DNA strand is lost by a cosmic ray, replicating or repair enzymes may not copy a template strand—and a mutation is born that is incorporated into the gene, altering its code. Chance evolution that confers an advantage—better communication that leads to honing of survival skills (as the product of mutations in the DNA of an ancestor)—can change the entire course of a species.

In the case of *Homo sapiens,* this evolution probably involved a mutation of the DNA involved in cerebral development of the left hemisphere, in the area of the *angular gyrus.* The descent of the supralaryngeal track in man occurred over a million years ago. With the development of mid brain connections to the *prefrontal cortex* in the last two million years or less, connections to the *Occipital lobe,* the *Parietal lobe,* and the motor areas responsible for lip and tongue movement developed along the *arcuate fasciculus.* Finally, development of conceptual sound-images could be translated into articulated sounds or speech, and comprehended as human language, with origins as old as bird-song.

In the theatre, when we perform onstage in our primary consciousness we are recalling through gesture and movement a range of responses that date back to these proto-reptilian epochs, and coordinated (or fine tuned) by language that is keyed by *angular gyrus* "secondary association"[18] responses of the left *temporal lobe.* Performing onstage through our traditions and customs, secondary consciousness draws upon the full repertoires of *cerebral cortex* mechanisms, and the global mapping of multiple reentrant local maps that are able to interact with non-mapped parts of the brain to express speech patterns as language. Movement gestures, remnants of ancient male plumage displays, are coded for tribal or national or universal significations. Seen through screens of consciousness, this doubling effect creates the appearance of personality.

In the 1990s, Performance Studies offered an intellectual way to assimilate secondary consciousness into the equation. The link to primary consciousness was less secure. The vehicles available to translate primary consciousness were themselves the regulatory product of a process that can only be discontinuously known. Put another way, the dramatic moment entails a full circle of fine tuned inferences and circumlocutions that originated synergistically within the limbic system, interacting with the *hypothalamus,* the proto-reptilian *brain stem,* the *cerebellum,* the *hippocampus,* and the *basal ganglia,* and all reflexive with the cerebral cortex. To focus actor training or stage production on cerebral processes is to focus on secondary implications that have little to do with the core of theatrical expression. In the 1930s, Antonin Artaud was not far from the mark when he suggested that, in his theatre, we "will seek to reach the mind by way of the organs, of all the organs, in all degrees of intensity, and in all directions."[19]

In summary, we owe our dominance of the earth to discontinuous neural maps in our brains, which undoubtedly spawned our capacity for change in the face of supremely difficult natural disasters: five cataclysmic asteroid encounters and an ongoing winter that has lasted to date one hundred and eighteen thousand years—and which may well have hindered Neandertal man, or at least hastened his disappearance.

The next time you cross the threshold of the stage to press the limits of the envelope of human presence, displaying Woyzeck's visions, playing out Segismundo's dreams, or portraying Zeami's ghosts, keep the limitless promises of the past in mind. The envelope of consciousness, from which to draw resources, extends back to the moment of the Big Bang, or a series of Big Bangs, and second generation stars, from which we have little hope of reaching around to observe any other measure of time/space.

PARAMETERS, TYPOLOGIES, AND PERTURBATIONS

At the beginning of this century, two schools of thought brought the new discipline of theatre studies into sharp focus. Erika Fischer-Lichte has noted that the German scholar, Wilhelm, proposed an end to the positivist school of theatre, based on principles of natural science at Antoine's *Théâtre Libre*.

In its place, Dilthey advocated that scholars in the field of humanities "should," in Fischer-Lichte's words, "focus on the individual work, which can only be understood by experiencing it."[20] At the same time, other scholars suggested that it was futile to write a history of the experience of an art form that was evanescent in the extreme. The only valid approach must be to continue to collect and present materials that would reconstruct the past performance. Hence the academic fervor that called for an examination of individual performances arose and fell on a new ebb tide that lasted through the nineteen-fifties.

This debate between theatre historiography and performance analysis resurfaced in the nineteen-seventies and nineteen-eighties with renewed demands for a more authentic analysis of performance, and with particular reference to contemporary performance. Semiotics was the vehicle of choice, and a system of signs that began with the Prague School in the nineteen-thirties was revived in the late decades of the twentieth-century. Other approaches were proposed including French structuralism, Russian formalism, the new criticism, the new historicism, materialist criticism, Lacanian theory, feminism, and systems based on significant figures:

Marx, Foucault, Derrida, Lacan, Braudel, Kristeva, and others, but each failed to produce a total system of description and analysis. However, semioticians like Marvin Carlson did offer a way out of the dilemma. Noting that semiotics, having focussed on the icon as the central element of staged performance, had overlooked the importance of the index as a key element, Carlson offered some amendments to the semiotic canon.

Since an icon is a particular kind of sign that resembles onstage what it stands for in real life, no representation could be complete without indexical relations of the characters' fictive life offstage also being present. For Carlson, the logical goal of this indexical element, in the form of letters, offstage dinner parties, foghorns, lighting effects, could now be expanded to include the spectators' offstage reality. As Carlson summarized, these indexical moments that unite everyday offstage space and theatrical offstage space "can be effected through a change of attitude that involves...seeing our lives as theatrical enactments, ourselves as characters in our own stage managing of these enactments."[21]

No more appropriate introduction to the indexical qualities of man as a theatrical species could be imagined. This text is based on indexical elements of the natural world as they effect fictive indexical qualities of the stage, a subject that has been explored in the living theatre for decades by such artists as Samuel Beckett, Karen Findlay, Laurie Anderson and Robert Wilson. Both offstage and onstage fictive indices have their sources in evolutionary, neuro-anatomical and cultural pressures.

However, conventional research cannot explain the internal complexity and self-regulating mechanisms of living organisms. Newtonian mathematics can not predict natural events in theatrical settings, or even the flight of a baseball in a stadium. Only certain laws of physics and mathematics in rigidly controlled settings can be "explained." But what of an actor's response to an empty house versus a full, over-flowing house, or a hundred other details of performance, that may change on any given night? Theatre as a complex adaptive system provides the only means of description and analysis of the discontinuous dimensions of the theatrical quality of *Homo sapiens.*

There are several large *caveats* to Carlson's invitation for indexical theatre space "to participate in the spectator's mental universe." In his commentary on fictive space of stage characters, Charles Lyons pointed the way to spatial orientation and concentration. For Lyons, "The spectator must be made aware of materials 'that will generate substantive images that will inform [the character's] consciousness."[22] Carlson credited Lyons with this insight, and went on to expand the boundaries of the indexical sign. However, by taking Carlson's dictate that we examine the spectator's world through character's speeches, a new set of problems arises. By confining that world to the interpretation of a stage event and insisting that the reality of the stage event is dependent on the indexical components of the spectator's mind, we arrive at theatrical consciousness and spaces of empowerment.[23]

The substantive images that inform this theatrical consciousness also originate in the fictive onstage and offstage indices, which reflect in turn indices in the reality of the external world. But attempts to enter the events of the spectators after they leave the theatre only intrude on the stage-action in disconcerting ways; one might as well speak of Renaissance characters in fictive space ordering your dinner or calling your stockbroker to change an allocation of mutual funds.

What is crucial to the analysis of indexical typologies is the participatory relations of the fictional mind of the character and the space that surrounds his or her total activities, as they cosume our mental frame in the presence of staged performance. The focus should remain on indexical components of man as an evolutionary species, with a limbic system that works a certain way, with a uniquely adaptable and discontinuous consciousness that allows an image to stand for a thing in the external world or in a fictive world. The same arguments can be made for biological rhythms and cultural patterns that create the very typologies we respond to, in human ways that trigger an empathetic response.

Having traced the core of indexical biology, evolution, and neuroscience in the theatre to semiotic theory, we are in much the same situation as the Prague Structuralists. A descriptive analysis of an event with so many discontinuous possibilities cannot address the performative moment. However, by acknowledging the nature of chaos in natural habitats, and developing a theory of complex adaptive systems to that event, the unpredictable becomes a part of the solution.

A semiotic description of the onstage and offstage fictive elements, without a leap of faith, will not lead to consciousness of the character nor to a dynamic description of the mechanics of stage productions to which we respond. The reason for this is that the symbol is anchored in consciousness, and the power of theatre—the sign—is deep in the neural mapping of the brain, deeper still in the evolutionary development of the species, much deeper than cultural consciousness.

Computer simulations of complex adaptive systems, on the contrary, are wholly in the realm of the new mathematics, or the new biology, and completely accessible to the idea of theatre as a series of discontinuous transactions by the mind and body of one individual toward another. Thus, the lock on contemporary dramatic theory, provided by semiotics through an indexical space analogy, is joined to the lock on contemporary mathematical theory, biological, cultural and evolutionary problems underlying all theatrical presentations. In its analysis and comprehension of the fictive world of the stage, complex adaptive systems can provide a methodology for an entirely new generation of civilization models in the natural world.

Chapter Seven

❦

Nature's Return

An intriguing new aspect of adaptive systems appeared in the summer of 1995, with Stuart Kauffman's text on self-organization as a basic principle of nature, on earth as in the cosmos. Kauffman notes: "What we are only now discovering is that the range of spontaneous order in nature is enormously greater than supposed."[1] Looking at the origins of life on earth, for example, Kauffman speculated that complexity itself triggers self-organization:

> If enough different molecules, for example, pass a certain threshold of complexity, they begin to self-organize into a new entity, in this case a living cell.

This replication of creation counters what I have previously described as the vastly improbable nature of our existence as a species, given the options.[2] Quite possibly, the self-organization principles of evolutionary biology that led to modern man were at work all along, creating diversity and high intelligence precisely because we are the most adaptive species on earth.

Life arose, argues Kauffman, not from accidents in the bio-evolutionary history of this planet, but from an expected fulfillment of the natural order. A large measure of that adaptability can be seen in the subject matter of this text. The creation of theatre is an infinitely probable componential strata of complex adaptive systems of human culture, evolutionary biology, and neuro-anatomical functions. Of even more consequence, I would argue, "is it 'natural' that earth has been hit by five galactic rocks?" Why not a dozen, a million, or none?[3] It is also feasible that—on the potter's wheel of creation—it takes five hits to shape a human species in the course of seven billion years, and everything is precisely on track.

Complex adaptive systems (CAS) are a new, discontinuous *Homo Ludens* paradigm, raised to universal status. Similarly, a theory of theatre based on complex adaptive systems can accommodate most, if not all, intellectual concerns. CAS are a form of play that can include all forms of seriousness, whereas seriousness, by its nature, seeks to exclude play.

Many perspectives in dramatic theory look at performance from a constantly evolving array of perspectives: structuralism, semiotics, feminism, deconstruction-

ism, perspectivism, phenomenalism, anthropological and neo-Darwinian creation-
ism, symbolism, rayonism, suprematism, intuitionism, neo-primitivism, as well as
the familiar futurism, expressionism, dadaism, and surrealism of more mainstream
art world forays. Only a modest assortment of contemporary artistic mind-game
hegemonies is listed. And yet, despite the ever-growing choices of accommodations
of the world of art to the world of the intellect, these scientific bases for literary discovery all involve a single goal, expressed and unacknowledged: an intention to define the unknowable relations between the artist and his or her occasion. The
necessity of defining the field by arbitrary means in order to establish a scientific
basis has, in the past, resulted in either a specialized language and a deliberate isolation of the dramatic event from everyday life, or an hypothesis that opens up a dialogue but does not deal with the processes.

Typically, theatre is seen as an external, encultured phenomena. One can appreciate the distinctions that Oscar Brockett makes when he describes *theatre* as a
form of art and entertainment and the presence of *theatrical* or *performative elements*
in other activities.[4] But from a perspective that sees theatre as a primal function, a
bio-intentional act of *Homo sapiens*, "formally" arising only in the last three thousand years of the species' existence, the discussion of culture and civilization becomes an outgrowth of this innate biological, sub-autonomous development.

However, the playgoer, in multitudes over many generations, also comprises a
complex adaptive system that energizes and transforms performance in various
ways. Even the "infinitely probable componential strata of plays, players, and productions within a culture and in a given period" constitute their own complex adaptive systems as well.

A THEATRICAL SPECIES

Theatre is one of the basic measures of human expression. We do theatre because
we are a theatrical species. In our Newtonian universe of the last three hundred
years, theatre has become a cerebral art form. There is a logic of sense memory, of
biomechanics, of learned techniques from the past.

Cerebral processes are, in themselves, too limiting. Rather than focusing on rational, connected behavior patterns, it is much more profitable to view theatre as a
discontinuous process of complex adaptive systems. Performance is as accessible as
the new biology, trade balances, studies of ecosystems, functions of the immune system, neuronal activity of the human brain, and even the distribution of goods and
services to a metropolitan area like Manhatten.

To reinvent a linear science of theatre, as did Antoine at the Théâtre-Libre in
the eighteen-seventies, is to distort the very nature of the beast. Theatre connects us
to our selfs, to ourselves, to our heritage from tiny forest creatures on a primeval
forest floor some one hundred plus million years ago, to perpetrators of the *Homo*

sapiens neandertalensis holocaust some twenty-eight thousand years before the present.

In order to rediscover the "book of nature," we might do well to note what Susan Sontag (in discussing "literature,") described as "a kind of calling, even a kind of salvation:

> It connects me with an enterprise that is over 2,000 years old. What do we have from the past? Art and thought. That's what lasts. That's what continues to feed people and give them an idea of something better. A better state of one's feelings or simply the idea of silence in one's self that allows one to think or to feel. Which to me is the same."[5]

Sontag's perspective is more narrowly focussed on the last two thousand years. I suggest a time span of interest of at least two and a half million years, with a specific focus on the last one hundred and twenty thousand years. Only when we begin to examine the *processes* of enculturation across many millennia, uncovered in a theatre performance, will we begin to understand the nature of the "idea of silence within one's self," achieved so rarely, sought so assiduously. Sontag presciently intuited the right church; she sold short the deep pew of an ancient brain. Attempts to objectify performance—and I speak of the early valiant years of semiology in particular—are as futile as efforts to draw conclusions from a focus on stringed puppets.

This text is predicated on the belief that we can find a process that is antecedent and common to the idea of theatre and the practice of theatre arts. We can then discuss what is taking place onstage as it is taking place in the mind of the spectator, without abstraction or special instance, or exception, or epistemological extrusions.

With the intense interest in all things scientific in the latter half of the nineteenth century, theatre attempted to put art and science into the same mold. During this period, the entire range of psychoanalyses was being uncovered and claims for its power were foisted on the world populace for generations. Theatre also adopted this practice, and dramas of the first three decades are rife with references to psychological nightmares that would have made Euripides proud.

But after the 1920s, and the extension of science into art domains under the headings of futurism, expressionism in particular, and the continuing dominance of naturalism and social realism on respective sides of the century-long conflict in ideologies, very little new ground was broken. With the rise of Fascism and Communism totalitarian movements in Europe through the 1930s, Western art retrenched. Much of the art that subsequently emerged was recruited in the services of specific regime's propaganda machines—East and West. Modernism played out in the battle between ideologies that dominated the 20th century.

It was not until an obscure paper on meteorology appeared, published by Edward Lorenz in 1964, stirring a new way of looking at the world of natural phenomena, that the sanctity of modernism was shattered.[6] This extraordinarily quiet revolution from a most unexpected place in the scientific community eventually spread to the art community. "Sensitive dependence" were the new buzz words for an in-crowd. Discontinuity of images in theatrical productions and in the world of scientific nature, became a tenet of the new norm. The development of chaos theory out of the new meteorology, the new biology in particular and the new mathematics, and of such innovative research projects as Swarm simulation systems, used to study neural networks and genetic algorithms, also signaled the end of the dominance of Newtonian physics. These scientific revolutions allowed social scientists and artists to apply the basis of chaos to everyday life in a discipline that became known as "chaotics," a term now credited to Keir Elam.

Edward Lorenz and his Attractor model, with its deterministic nonperiodic flow, and his conjecture that all natural systems are governed by "initial conditions," truly began the postmodern era. If history is any indication of the future, we will play out these theorems in the theatre until the year 2030, develop permutations on them until the 2080s, and then begin anew to decipher the sense of human presence onstage. For the last several hundred years this procedure has been a signature behavior of the human psyche in its communal form.

Semiotics, the one bright spot in the 1930s, lay stillborn after the nineteen-forties despite valiant efforts to revive it and apply it to theatrical production in the nineteen-seventies, nineteen-eighties and early nineteen-nineties. All efforts failed for one reason: semiotics is a descriptive mode and not an active mode of interpretation. Similarly, deconstruction criticism in the nineteen-eighties, which is little more than the annihilation of cerebral processes of interpretation, leaves off where neurodrama begins! Hence, without the first half of the curve, the second half would make no sense. And without the second half, broadly sketched here in only a few instances, the movement into the twenty-first century would likewise have no continuity.

However, since discontinuity is one of the features of the newest brave world, one should not despair of the failure of any linear flow of stage representations. The newly articulated deep structure of the mind can incorporate theatrical presence of the past with theoretical discussions of the theatre of yesterday, today and any morrow. These structures have always been there, and have always been utilized by artists, acknowledged or not. Ultimately, the goal is to provide both performers and theoreticians with a common basis of just exactly what it is that we speak of when we discuss the thought, character, action, gesture, intention, and dialogue of any element of performance *at the moment of its communicative fullness*, and not after the fact—in the isolation of the study or the classroom, or even in the separateness of the actor's dressing room.

As I look at the development patterns of a three-year-old child, at an age where the left hemisphere begins to dominate, the theatrical instinct is developed. Why? Because the right hemisphere is using image formation and perceptual shapes that are communicated to the outside world by means of the left hemisphere, and the *angular gyrus* secondary association areas that can now explode with information. The sense of theatre is created by the absence of direct communication of those spatial, creative sound images in the right hemisphere, the absence of a direct mediating vehicle for the right *angular gyrus*. Were both right and left *angular gyri* present, the nature of theatre would be intolerably boring to at least one half of the functional brain. This proposed binary species is of course, non-existent. At some millennium in the distant future, our species may develop that capacity, as surely as *Homo sapiens* developed a capacity for reflection of the highest order some 40,000 years ago.

This new age might regard theatre, born 40,000 years ago, as an interesting transition point, where secondary association areas of the left *angular gyrus* of the temporal lobule of the left hemisphere, stood alone. "A quaint age," some might say, "whose earliest inhabitants thought they were speaking to god, and heard the voice of god, when he or she listened to the voices in their heads," (emanating from the right cerebral hemisphere and not from mental instability!); voices which arose from a source that could not be identified.[7] What can we do with a right-hemisphere centered culture that feels at one with the land, and creates music of the here and now *Dreamtime*?

The questions that immediately arise include: "What of the acts prior to *Gilgamesh*, for example, Where are the plays of modern man at the edge of inspiration, or at the dawn of consciousness?" And "Why are there no texts to accompany the dying hunter/victim and the bison in Lascaux of south-western France or in Altamira, circa 29,000 years ago. If there is no written tradition, was there an oral tradition?" One must argue for an incredibly major functional change in man's perceptive abilities in the generations just prior to circa 40,000 years ago. Is this time period linked to the disappearance of *Homo sapiens neandertalensis*, and the rise of modern man as we know him today, the theatrical man with language and gesture and formulations, the man with reflection?

At the end of the Ice Age, circa 12,000 years ago, *Homo sapiens* emerged with all the capacities of our consciousness. Within a few thousand years, languages appeared, domestication and presumably theatrical activities were enjoyed. However, no one can say for a certainty that entertainment was a civilized value or that civilization, as we understand it today, existed. The only explanation for a radical change in the history of a species that is over four and a half million years old is that an incredible breakthrough took place in the brain of the species. The breakthrough came, without question, in neural connections to the *pre-frontal lobes* and the secondary association areas in the left *angular gyrus* of the species' left temporal lobe, which permitted the development of a conceptual frame for language that included visual stimuli. No endocasts exist for proof of this hypothesis, and speculation must continue. But there is no other explanation why the species should advance exponentially in a few thousand years, where it had remained essentially unchanged for several million years.

The great leap forward came as a result of *frontal lobe* connections to these *secondary association* areas that could link the triune brain to a reflective species via the association pathways. Unfortunately, the hard-wired creature that emerged was as much a judgmental as a compassionate creature. Divisions between the civilizations that arose have continued to this day primarily because the power to theatricalize is also the power to enslave. Until the quadrune brain emerges, if ever, in a parallel development of *association areas* of the *angular gyrus* in the right hemisphere, the altruistic compassion may never dominate. Conversely, the power of man as a theatrical species may be seen at that moment as a passing phase, much as the image of god for our early modern man, seen as the voice of the inner self, so frightened its possessors.

If a quadrune brain finally emerges, the frightening space within—site of the beast that was "the forest through which man tracked his own awareness," as Robert Wilson described it in *The Forest*—may be gleaned as another early expression of those terrors Artaud, Nietzsche, and the unknown author of *Gilgamesh*—endured across 4300 years.

COMPUTATIONAL STRATEGIES FOR THEATRICAL CONSCIOUSNESS

Despite cultural and social differences in Western and non-Western civilizations, the nature of theatre, and particularly the preparation for performance of actors of Western and non-Western countries, appears from direct observations and discussions to have a common root. A Malaysian actor and an American actor express the same emotions, with only the cultural conventions fine tuning their specific and disparate responses. This drawing out from within of energies and inspirations brings all to the mystery, the chaos, of limbic responses—of fight, flight, sexual encounter and hunger. In attempting to analyze this interiority, Chaotics, the New Theatricalism and the general theory of complex adaptive systems expand exponentially the resources and the disciplines that are relevant to theatre studies.

In examining these new relations and in employing their significations in the theatre—much as one might examine Foucault's "geological strata" of social relations—a series of social, cultural and biological relations emerge—both historical and contemporary. Looking at Foucault's stratas, chaos theory in theatrical performances might be judged to be a logical development. Set in motion, initial disturbances are the origins of vast changes in the communicative gestures. But that is only a tiny portion of the problem. For character to be alive onstage, responses must seem to come from the actions taken. Chaos theory can begin to explain this phenomena. But for responses to be truthful or authentic, there must be continual development and changes in the internal regulations of the character. That "internal discontinuous change" can only be explained by a theory of complex adaptive systems. Hence, the nature of theatre is intimately dependent on the nature of mind and the human brain, which is arguably the most complex adaptive system in the known universe. And while complexity is usually seen as a series of differential mathematical equations of physical systems that explain how humans interpret with their minds objective phenomena, theatre continues to bear a special relationship to complexity precisely because theatre is the human brain functioning at its highest level.

At some point we will learn to dissect the processing human brain, perhaps by using functional Magnetic Resonance Imaging techniques to extract sound images, and then to analyze these sound images by employing techniques derived from theoretical physics, such as computational mechanics of cellular processes.[8] Research into these areas in the last few years has led to exciting prospects. But concrete applications for this very young science are possibly another decade away. For example, how to record brain activation without invasive techniques, or to correlate in real time the dynamic state of cellular mechanics of human brain activation, has been debated. George Cowan addresses current fMRI research as follows:

> As for fMRI, it measures blood flow and is associated with neuronal activity because synaptic discharges are major consumers of energy. This is supplied by cerebral blood flow. Blood carries glucose and oxygen to working neurons on demand, and so fMRI measures the degree of local activity of neurons. These measurements are translated into more or less highly resolved images with appropriate software.[9]

The program in the physical sciences is now firmly rooted in mathematical equations for computer simulations. Practical applications of complex adaptive systems are becoming prevalent everywhere, from ecosystem functions to Aids research, to analysis of vehicular traffic patterns, to neuro-anatomical investigations of the nature of the thinking human brain. For the arts, the issue has not yet been acknowledged. Complexity is still the field of artists of extraordinary power, whether it is Pablo Picasso, describing his paintings as "cosa mentale" or Samuel Beckett, acknowledging his writings as the fruits of a "sensibility." In defining the neuroanatomy of modern man, for example, the newly discovered role of the *thalamus* and the *basal ganglia* in the old *neoreptilian* brain of man as the arbiter of consciousness has reestablished our link with ourselves. Suddenly, the Nietzschean abyss seems much more navigable.

In theatrical texts, we can track at a point in history the discontinuous biological expression of the species in social and cultural moments; in performances of the present, we can examine the history of conventions that carried those same social and cultural elements of biological imperatives to the present day. Somewhere along a range of differences, a solution or an interpretation of a performance moment—

seen in its widest possible contexts (ranging from shamanism to post-colonial theory to contemporary performance to dance to food diet to public demoralization)—may be apparent. But, only for a given moment: the relations are succeeded by new, discontinuous information. Within that range is the power and transcendent prospect of complex adaptive performance that draws from a core of evolutionary change of cellular life forms of almost four billion years.

Pure research on neuroanatomy and convincing neurobiological evidence on the nature of perception has been published widely in recent years. For students of a theatre of images, for example, or others who are generally interested in the emotional context of theatrical gestures and split screen projections, or multiple frame foci, this research is illuminating. Robert Wilson in particular has dealt with the divisibility of consciousness phenomena for decades. Using split screens, tape loops and projections of images on numerous scrims in the same scenes, he invites his viewers to find their own interpretations of the dramatic moments that are, by their very nature, discontinuous. Like electrons moving about a magnetic field, (created by the spectator's consciousness of the nature of the fictive space he or she is observing), some screens collide, some scrims remain isolated, some sound bites from the tape loops change form: in an instant. A new stage image is created, as the typologies—genetic, cultural and neural—rearrange the spectators' consciousness.

How is this possible? Striking experiments in aberrant communication pathways by such researchers as Norman Geschwind at Harvard University, and Albert Galaburda at Cornell University Medical Center indicated the functioning neuroanatomy that has always been at the base of Wilson's art. Patients with *commissurectomies*, split brain procedures in which their *corpus callosums* have been severed, either as a result of stroke or surgery for otherwise uncontrollable epileptic seizures, exhibited the ability to name an object seen on a screen only after it is drawn by their left hand, for example, or reveal the ability to write a complete sentence that they are utterly unable to read.

In the "naming of the object research," a split brain patient was asked to look at a dot on a screen. An image was flashed to the right of the dot; the patient was able to name this image, since the image on the right was transmitted to the *visual cortex* of the left *hemisphere*, where it also connected to the site of language in the *angular gyrus* of the left *temporal lobe*. However, when the image was flashed to the left of the dot, the test subject was unable to name it. The explanation offered by Geschwind was quite simple. With a severed *corpus callosum*, the information from the right *temporal lobe* stayed in the right *hemisphere*. But when she was asked to close her eyes and draw the image with her left hand, also connected to the right *hemisphere*, she was able to do so. And, having drawn the image, she was then able to name it.

Further Geschwind research into a severed *arcuate fasciculus*, connecting *Broca's area* with *Wernicke's area* in the left *cerebral cortex hemisphere*, discovered a patient who could take dictation, but who was utterly unable to read what he had written. Explorations of this kind were common in the early theatre work of Robert Wilson, who would feature a brain damaged young man as the inspiration and living actor/character of *Deafman Glance*. This research is not to suggest that Wilson's theatre appeals to those with severed *hemispheres*, but rather, that the processes of comprehending language and visual stimuli are discrete and separable discoveries in theatre research and in neuro-anatomical exploration by medical scientists.

This "divisibility of consciousness" describes an arc between perceptual consciousness (based on non-referential neural processes of which the spectator may not be aware), and intellectual consciousness (based on semantic information about the stage stimulus) is clearly the domain of post modernist artists like Richard Foreman, Robert Wilson and Laurie Anderson. From da Vinci to Picasso, *cosa mentale* has always described the processes. But no one prior to the post modernists—and I include Samuel Beckett in this grouping—made a virtue of the very separable elements of artistic consciousness. (Pirandello separated plot, character, thought, aesthetic distance.) The discontinuous nature of image and sound, language and context, perceptual and intellectual consciousness, underlies every aspect of theatre and the visual arts. The fact that hard science is also fully engaged in exploring the limits of complex adaptive systems that reinforce current research in the arts and the humanities is of great comfort. It was not always so.

Since the seventeenth century, epistemological studies have dismissed subjectivity as irrelevant. For three hundred years, the Cartesian mind-body duality inherited from Aristotle posed insoluble problems in objective science. However, the basic problem is to define the notion of an objective event in real time. As Husserl and Heidegger might have argued in infinitely more subtle terms, a hermeneutic of non-reducible subjectivities at the level of consciousness is the only posture acceptable for the interpretation of human sciences and the arts. The argument presented here at its most basic is that subjectivities may be reduced to infinite measures of objective data—to Zeno's grains of sand—but the interpretation of artistic works lies at the level of non-reducible, indexical offstage subjectivities of consciousness. These subjectivities appear to be of a parallel rather than of a serial nature. Hence the limitations on artificial intelligence as it was envisioned in the 1960s, generated in the 1970s and 1980s. More recently, the concept of complex adaptive systems has emerged as a method by which to understand those fundamental indexical processes that shape almost every aspect of human life—from computer viruses to Dow Jones averages to past and current styles of theatrical performance.

The nature of theatre is defined by the nature of evolution. Formal observation of training sessions of Western and non-Western actors, and readings in cultural studies with multiple connotations of performance, suggest that the creation of forms of formal theatre performance arise from a common source. The quest /drawn from that sampling that would return us to the Foucault/Heisenberg dilemma of measuring and comparing one sample from the past with another sample from the "present." The goal of this computational mechanics would now be to define series' limits, relations, strata, over periods of time and to individualize different strata, to juxtapose fMRI signals from nested neural networks without reducing them to linear schema. And what is "the present" against which we measure another time-space? What neural signal, against what other inner criteria? And how is the maintenance of a steady state to be deciphered and measure against an increased signalling that indicates intentional activities? At the very best we have assayable data. At the very least we have patterns of relationships which might, in all their ambiguity and discontinuity, provide analysis of individual

QUADRUNE THEATRICAL BRAIN OF *HOMO SAPIENS*

The history of species indicates that catastrophe creates the momentum for change, and not abundance. In times of ecological or climatic crises, hard-wired species disappear. Adaptability is the key to survival, particularly in the hominid species. In times of climatic crises, for example, as species are faced with extinction, the case for punctuated evolution can be made. In isolation in the last great Ice Age, bands of *Homo Sapiens* species enhanced their survival with new and better communication skills. Vocal apparatus emerged as a biogenetic accident, and the modern supra-laryngeal vocal tract emerged as early as 190,000 years ago.[10] The *secondary association areas* of the left *temporal lobe* also began to function, and the *angular gyrus* emerged as a speech center for modern man, all connected to the *prefrontal* "executive brain" *lobes*.

The argument that increased resources allows divergent variants of the species to emerge fails when the question of what abundant resources were available 190,000 years ago, at the beginning of the great modern ice ages. Instead, climatic crisis set the stage for isolation and support of novelty, where that novelty helped the species to care more effectively for kin.

Four million years ago, the disappearance of heavily forested cover in Africa, created the savannas that forced *Australopithecine* hominids to adapt to open country. The result was *Homo habilis*, early man upright and walking on two legs, with the front legs now used as arms, and cranial capacities to develop into *Homo erectus*, the handy tool-making, savanna-exploring man. Nothing in the literature of evolution supports the theory that plenitude leads to diversity. Species seek the lowest, most efficient form of life. In times of hardship, the most adaptable species seek new ways once again to reduce the stress of living to its least intrusive level.

One would suggest that, with the development of the left *angular gyrus* in the *temporal lobe* of *Homo sapiens*, the next major step might well be the development of a right *temporal lobe angular gyrus*, to set up two thinking communication centers in man, the extant left lobe consisting of a rational center for communicative interpretation for physical manipulation of the external world, and the right, a center for expression of the interiority, an imaginative, reptilian-organized center for communication.

That right hemisphere center, linked by the *corpus callosum* to the left center and association mapping of neuro-anatomical areas, might not come for another hundred thousand years. When it does arrive, as a result of some catastrophic occurrence such as the destruction of civilizations by man-made ordinances, it might be necessary to redefine the species. We are from a holistic reptilian base. Can we subdue the rational aggression centers in the left hemisphere that have been the heritage of 4.6 million years of our theatrical species, as we know it today?

LIMBIC GAMES

If theatre is a dynamic enterprise, whose agents cannot be measured in a linear pattern, the closest approximation to the discontinuous typologies of this art form must come from simulations.

The most practical simulations are those that can be adapted to computer programs, where millions of simulations may be run in a few minutes, rather than several over the course of a thousand years! Can theatre as a complex adaptive system be represented as a computer simulation?

The issue of developing adaptive controls for complex systems is, by its very nature, arbitrary. Theatre's imitative properties are natural; its communicative elements are natural, but the interpretation of theatrical images and symbols by cerebral processes are human impositions on the natural order. In computer modeling, this difficulty remains paramount for any given formulation. The investigator may be reduced to admitting, as Seth Lloyd noted in "Learning How to Control Complex Systems," that intuition is frequently the best resource.[11] Usually the computer programmer "does not even know the dynamics of the system…enough to program a computer to simulate them." Using the model of landing an airplane in real time, Lloyd summarized the difficulties:

> Even though the computer's predictions are accurate, and the effect of a possible sequence of controls can be determined in a short time, there are an exponentially large number of sequences of possible controls, and to verify the consequences of each one of these sequences in a search for the optimum sequence takes exponential time.[12]

For our purposes, not only is there difficulty with computer modeling, there are the indeterminate factors of consciousness and interpretation—of perturbation or "noise"—that cannot be eliminated from real time analysis, even using adaptive evolution, adaptive culture and adaptive neuroscience as the dynamic bases. The possible formulations for theatre as a CAS may be incalculable. For that reason, intuition and the ability to choose a minimal set of internal models, diverse agents and nonlinear quotients, is crucial.

Recent work on screening devices of sound waves, using computational mechanics from theoretical physics formulations can be explored in computer simulations. Such combinations of internal models as outlined by Norman Holland
require not only a qualitative picture of the system's dynamics but also a model that
emulates the system's methods of processing information—a parallel duplicating
event that becomes the computer programmer's principal goal.

In order to formulate a computational base of theatre, one combination of internal models (the actor's psyche), diversity of agents (cast of a show), and nonlinear
quotients (indexical culture, evolution, and neuroanatomy in the fictive stage
space), can serve as a basis. Holland's second criteria, that of selecting both a system
that carries with it a built-in symbol-system and a performance system that speci-
fies the agent's abilities in the absence of further learning, must also considered.
Holland's final admonition to the experimenter: the ability to select insightfully an
inductive apparatus—a dramatic moment of character choice—"that modifies the
performance system as experience accumulates," is a lesson that every successful
playwright must learn! As noted from the beginning, man's ability to look at himself, and see in that reflection a glimpse of the universe and a fulfillment of the
natural order, is nowhere more complete than in theatrical performance.

The interpretation of symbol systems demands the interpreter reach into a
complex adaptive system of neural realms, evolutionary niches, and cultural mores.
In the theatre, those systems, niches and mores relate to emotions; concerns, governed by the *limbic* system. Given this situation, it might be suggested that the horizontal xy axis describes the predictable and deterministic features, while the vertical
z axis describes the behavior of any nonlinear dynamical systems. In the case of
theatre as a complex adaptive system, vertical information on neural complexity of
the actor and of the spectator, cultural complexity on the play's venue, play choice,
play's success, of performances of one style or many over "n" years, for example,
may form probabilistic information tables. Genetic typologies that look at the evolution of consciousness, of speech, of formal communication, within the species, are
vitally necessary before genuine computer modeling can begin. At that point the
speculation really begins, for it is here that all the unpredictability factors arise, the
"noise" factors of consciousness that cannot be eliminated from any naturally occurring dynamic.

In a dynamical systems theory in which any one potential solution can be
shown to be correct or incorrect, but the space of potential solutions is exponentially large and unverifiable, it may be necessary to introduce a framework that unifies the description of regular and random features.

1.   The three theatrical *typologies* are cultural, genetic, and neural adaptive complex sys-
     tems.
2.   The *controlling scale*. The model that the theorist possesses for the system in order to
     construct algorithms for this complex nonlinear system in the presence of incomplete,

apparently random behavior, is our *limbic system,* comprised of four *limbic system* elemental responses: 1. Fight; 2. flight; 3. sexual encounter; 4. hunger

3. The *set of rules* are those invented by the playwright, or the actor, the director, and the design team. In the case of human beings, the critical faculty provided by *reflection,* a discontinuous neural event that separates our species from otherwise hard-wired species, has been instrumental in our ascendancy to power over physically more impressive creatures. Reflection determines the plan of attack: artistic, intellectual, societal, biological, actions that will shorten the odds of success. With that physical pattern in place, the apparatus of theatre can spring from the mind (now!) of our ancestors and the neural mapping begin in earnest.

4. The interpretation of unpredictable and apparently random behavior can be identified with the socio-cultural, indexical qualities as they are played out, and which constitute part of the probabilistic interpretation of the game.

These combinations of internal models, diverse agents and nonlinear quotients end all similarities with traditional Newtonian-based models of hypothetical nature, particularly in respect to the evolving diversity of agent functions in complex adaptive systems. As regards change, where there is a weakness, there is, potentially in the long link of historical events, a problem that will modify or expel the species from further development. On occasion, there is strength that spells inflexibility in adapting to new environmental conditions or cultural requirements.

Complexity, it bears repeating, is a natural and internal process, expressed in its most concrete way through the theatrical nature of the species precisely because of an adaptive neural discontinuity that has enabled us to survive 4.6 million years of change, within a phylum that began 195 million years ago, on a planet that took seven and a half billion years to evolve, with the assistance of at least five space rocks.

If there is one thing that has separated our species from all others, *Australopithecus anamensis or A. afarensis* from *A. robustus* or *A. boisei,* for example, or *Homo sapiens* from *Homo sapiens neandertalensis,* it is our unwired flexibility for change. Reaching the stature almost of natural law, we might claim that our adaptability of complex cultural, biological and neurological systems to environmental and climatic dynamics wins out over strength and inflexibility *every time.* While most of the laws that govern genetics are known, the internal genetic modeling that takes place in man can never be absolute.

Furthermore, complexity does not lead to simplicity but to more complex solutions or options for solutions in the genetic chain as generations succeed another. Hence, the break in the Darwin paradigm is not complex adaptations that lead to clarity, but complexity that infinitely masks itself as comprehensible selection.

The improbable accident of the billion-year tape of genetic history turns out to be adaptability. Theatre, with its freedom to interpret discontinuous fictive events

according to arbitrary patterns, is the most unique expression of our species' adaptation to the visible and fictive worlds.

The improbable accident of the billion-year tape of genetic history turns out to
be adaptability. Theatre, with its freedom to interpret discontinuous fictive events
according to arbitrary patterns, is the most unique expression of our species' adaptation to the visible and fictive worlds.

The ability to understand *Gilgamesh*, the *Natyasastra*, the legends of Grecian
Olympian heroes, the traditions of the Spanish Golden Age, the tragedies and
comedies of the Elizabethan Age, the works of Chikamatsu, Shakespeare, Racine
and Beckett, and to function fully in the real world are essential. Each dramatic
moment is a product of that pre-historical record as much as an expression of the
indexical qualities of an art form that is inseparable from *Homo sapiens*.

In its broader sense as the conscious perception of human presence in nature,
theatre is a gift of freedom that our forebears, sixty-five million years ago, gave to
us to know the nature of complexity in a world that would otherwise seem arbitrary,
uninspired, leaden. In the history of species, it is a most curious fact that this gift
was not fully propagated until 80,000 years before the present, nor flaunted thereafter for 77 millenniums.

INDEXICAL DYNAMICS

The theatre has long been considered the most complex of all art forms because its
ingredients, many of which are arts in themselves, combine in discrete proportions
to create the theatrical event. The French semiologist Roland Barthes once described the nature of theatre as a "polyphony of signs ... a kind of cybernetic machine...emitting a certain number of messages."[13] What he had in mind were those
messages the spectators received from the settings, costumes, lighting, gestures and
speech—to name the most obvious signs. Semiotics, or the science of signs, is the
study of this interplay of icons; Marvin Carlson's adept description bears repeating:

> An icon is a particular type of sign which in some manner resembles what it stands for, as
> an outline of a cow on a highway marker warns motorists of possible real cows in the vicin-
> ity.[14]

More particularly, Carlson reviewed the nature of *iconicity*, and noted its starting
point in Jan Kott's 1969 statement that the basic icon of the theatre is the voice
and body of the actor.[15] To Kott's premise, there are exceptions: one need only
think of Filippo Marinetti's *Feet* (1915), where a half-raised curtain reveals "feet"
gavoting about the stage, or of Samuel Beckett's *Breath*, where the action consists of
a modulated cry echoing across a stage-space of detritus. Furthermore, these messages are only a portion of the theatrical polyphony.

It cannot be emphasized enough that the ingredients of "communicative signs"
of performance emanate from both sides of the proscenium. For passive spectator

signs, consider productions of *Romeo and Juliette* or of *Tartuffe*, to a full house, and then to an empty house; to a high school audience; and then to a gathering of senior citizens. For active spectator signs, consider any production by Julian Bond and Judith Malina's *Living Theatre* to the Sexual Freedom League in the 1970s or the San Francisco Mime Troupe's performances to inhabitants of Berkeley's Live Oak Park in the same decade.

In the 1970s and 1980s, theatre semiotics sought to convince the theatre discipline that, whether the topic was historiography or performance studies, semioticians could analyze the formal signs and codes of performance more thoroughly than any other approach. But the results were, and have been at the least, laconic. The failure stems in part from the impossibility of completing the task of creating a total system of description and analysis with signs that refuse to be contained within any objective frame. There is always something left over in any analysis, a quality embodied by the performer which is "born on the site of the Other."[16]

Have we reached a point where traditional concepts that date back to the Enlightenment belief that the intellect of man is the measure of all things must now give way to chaos theory and quantum physics complexity? If the answer is yes, then a genuine paradigm shift is underway.

Traditionally we might suggest that spectators, singly and as a group, communicated signs to the stage, to the actors, and to the actors-as-characters; depending on the environmental milieu, specific social and political codes, and even scientific and economic precepts dominant in the strata of culture present at any given performance. In Barthes's view, when the communicative signs exchanged between actor and character, character and company of actors, characters and spectators, are reciprocal, the nature of theatre becomes a reality. But what kind of reality is this privileged domain of the icon?

Carlson has pointed out that an important clue to the future direction of theatre study lies in the expanded view of fictive space of almost all theatre. What he had in mind were those signs that come from beyond the stage space, the indexical signs, as he described them, that give us the dimensions of the action, or the action in context.[17] I would suggest that the assumptions in Barthes' description of theatre, the assumptions of a catalog of theatrical elements, the assumptions of traditional analyses of culture, do not even begin to assay what theatre is, any more than the topography of a region's land mass or the geography of earth at certain latitudes indicates the nature of Russia, or Cambodia, or of Canada. What Carlson's analysis permitted was the fictive world beyond the stage to become a central concern in any semiotic analysis.

Hitherto restricted by the closet study of semioticians, the reality of fictive stage space demanded an inclusory and contextual assessment of indices from an offstage reality, whether it is "second door on the right" for Didi's bathroom chores or the twice-sounded breaking of a harp string in *The Cherry Orchard*.. Approaching the problem of identifying theatre space from the perspectives of the artist and the spectator, or the performer and the critic, or the designer and the theatre historiographer or performance theorist, created even greater problems. For the moment, in pursuing only the performer's contextualization of theatre space, Carlson cited Charles Lyons' perceptive view that the artists' rendering of character and the context of that character's representation was only completed when the spectator also participated in that contextual exercise.[18] Noting that this exercise in Lyons' analysis takes place onstage, Carlson carried the image to indexical signs beyond the immediacy of the visible space off the stage. A unified portrait of semiotic and indexical space was complete. But the question of exclusion from the designated frame, and the problem of whose frame was it, and why was it imposed on the reader, remained.

In a 1995 essay, Erika Fischer-Lichte summarized the situation very clearly:

> The object of the analysis is never the event itself. That is to say, even in terms of methodology, there is no difference between historical research and performance analysis.... It seems highly advisable to check the traditional differentiation between different fields of research, or at least, to become aware of the fact that this differentiation does not spring from any *a priori* reason but rather from the particular intentions placed on the agenda by certain schools as well as single researchers within a specific historical context.[19]

Fischer-Lichte dismissed the ground for difference between competing ideologies of the modernist school, and suggested that we re-examine those conditions given to the spectator from contemporary artists, or from artists who have pushed the envelope of contextuality decades before the present.

In his discussion of the function of foreign indices, Carlson was equally adamant on the need for change. Indexical signs can be directed at the fictional onstage and offstage realities of the consciousness of the character. By deconstructing the basis of semiology, his analysis provided an entry point to complex adaptive systems. But Carlson would push beyond that agenda. Noting that indexical signs (almost exclusively ignored by semiotic studies), can become more relevant to the total stage picture if those signs can be directed at the parallel real world of the audience, Carlson suggested that spectators see their lives "as theatrical enactments," ourselves as "characters in our own managing of these enactments." He continued:

> When the indices of theatre point not to its own offstage but to ours, they may point not only to general or particular places in our mental universe, but much more importantly to spaces of action and empowerment, where still unexplored possibilities beckon.[20]

The term "empowerment" suggests yet another intellectual concern, which would again limit the possibilities of an adaptive system frame, but Carlson's reference to the spectator's parallel offstage indices finally opened the door to a study of theatre as a complex adaptive system.

Originally applied to science, and a paradigmatic shift away from linear, reductionist, simple cause-effect models, the theory of complex adaptive systems was created to address problems in the world that can only be resolved by the functioning of internal adaptive agents and models to changing modalities. Situations that respond to complex adaptive formulations include such diverse processes as the pre-biotic chemical reactions that produced life on earth, biological evolution itself, the functioning of individual organisms and ecological communities, the operation of biological subsystems such as mammalian immune systems or human brains, aspects of human cultural evolution, and the like.[21]

Complex adaptive systems provide a framework for new science and for art in the next century. These modalities are shaped as agents in stage space, or as limbic functions within a stage character's consciousness of that perceived visible and extended stage space. These latter agents are a product of evolutionary, biological and cultural indices. Hence, theatre as a complex adaptive system involves all of these typologies, in a mix that is truly astronomical in scale, and yet simple for the human mind to comprehend.

Necessitating a series of discontinuous quantum leaps or exercises in the coding of "relativity indices" that no semiologist or theatre historiographer could ever compress to a typed page, complex adaptive systems work as much from intuition in the creation of models as from hard facts. In the best of cases, the hard facts are strained from evolution, biology and culture, and applied as computer simulations. No schools of thought or agendas can be included. As Fischer-Licht has tangentially pointed out so carefully, in the essential dynamic of this most evanescent of all art forms (that expresses the history of the human species as does no other indice), there are no other hard facts from which to proceed.

Chapter Eight

❧❀❧

The Discontinuous Cortex

EVOLUTION INTO COMPLEXITY

I think, if I can so soon judge, I shall be able to do some original work on Natural History.[1]

Charles Darwin

Charles Darwin sailed to the Galapagos Islands to find proof of the literal truth in the ordering of species, as proclaimed in the Christian Bible. In establishing a science of evolution, Darwin believed he had deconstructed specific Christian beliefs, Not until the discovery of Chaos and Complexity Theory in the 1960s was an adequate answer to his doubts available.

On February 10, 1832, at the beginning of his three-year sea voyage of discovery, Charles Darwin wrote those prescient words to his physician father, Dr. Robert Darwin. Darwin was to serve as naturalist on the vessel. Not the least of his duties was his assignment to provide evidence that would establish, once and for all, the literal truth of the story of creation found in the Old Testament.

On September 7, 1835, the H.M.S. Beagle left South America and sailed west for the Galapagos Islands. Darwin's diary included the following entries for September 26 and 27 respectively:

I have specimens from four of the larger Islands…. The specimens from Chatham and Albemarle Isd appear to be the same; but the other two are different. In each Isd each kind is exclusively found: habits of all are indistinguishable…. If there is the slightest foundation for these remarks the zoology of the Archipelago will be well worth examining; for such facts would undermine the stability of species.[2]

The following year Darwin returned home, convinced that natural selection was the key to evolution. His belief was confirmed by his reading of Malthus's theories on population: "It at once struck me that under these circumstances favorable ones tend to be preserved and unfavorable ones to be destroyed. The result of this would be the formation of new species."[3] In November 1859, after years of protean work during which he was hampered by consistent doubts that seriously affected his health, Darwin published his *Origin of Species*. The text included the following revelation, clearly contrary to accepted Christian doctrine:

It is a truly wonderful fact—the wonder of which we are apt to overlook from familiarity—
that all animals and all plants throughout all time and space should be related to each other
in groups, subordinate to groups, in the manner which we everywhere behold—namely, va-
rieties of the same species…. If species had been independently created, no explanation
would have been possible…but it is explained through inheritance and the complex action
of natural selection, entailing extinction and divergence of character.[4]

And, in words as modern as the anguished cries of Friedrich Nietzsche or of Samuel Beckett, Darwin raised his own protest against the prevalent conception of God:

A being so powerful and so full of knowledge as a God who could create the universe, is to
our finite minds omnipotent and omniscient, and it revolts our understanding to suppose
that his benevolence is not unbounded, for what advantage can there be in the sufferings of
millions of the lower animals throughout almost endless time.[5]

In Beckett's *How It Is*, the narrator's words echo this cry:

fallen in the mud from our mouths innumerable and ascending to where there is an ear a
mind to understand a means of noting a care for us the wish to note the curiosity to understand an ear to hear even ill these scraps of other scraps of an antique rigmarole[6]

One can suspect Beckett has Darwin in mind when Pim asks if he "might not put an
end without ceasing to maintain us in some kind of being without end and some
kind of justice without flaw." (139).

No one was better equipped to proclaim a new vision than Friedrich Wilhelm
Nietzsche (1844–1900). In *Thus Spoke Zarathustra*, Nietzsche proposed a new
morality:

Behold, I teach you the overman. The overman is the meaning of the earth…. Once the sin
against God was the greatest sin; but God died, and these sinners died with him. To sin
against the earth is now the most dreadful thing.[7]

During the last years of his life, Nietzsche transformed the agonies of his mind
into *Thus Spoke Zarathustra*. Within a period of what he claimed to be "ten absolutely clear and fresh January days" in 1883, the text was hurriedly written down.[8]
Rejecting Christianity as decadent and immoral, Nietzsche sought a superman who
would initiate a new heroic mortality. The passage from beast to overman was dangerous. But for those who risked the journey, Nietzsche promised great rewards: "A
new pride my ego taught me, and this I teach men; no longer to bury one's head in
the sand of heavenly things, but to bear it freely, on earthly head, which creates a
meaning for the earth."[9]

It might be conjectured that the death of God was the death of the right hemisphere as a site of godhead (Jaynes postulate). Nietzsche mixes Christian with pagan
beliefs. Furthermore, in the absence of complexity and chaos theory and relativity,
a century later, the German philosopher fills the void with a vague metaphoric reference to "earth."

This conscious creation of the Anti-Christ (Nietzsche himself) was an effort to break the morality of the age, and to prepare for a new era of scientific discoveries. In a letter to Jacob Burckhardt in September 1886, Nietzsche reaffirmed his dialectical approach to history:

> I know nobody who shares with me as many prepossessions as yourself: it seems to me that you are working on the same problems in a similar way, perhaps even more forcefully and deeply than I, because you are less loquacious.... The mysterious conditions of any growth in culture, that extremely dubious relation between what is called the "improvement" of man (or even "humanization") and the enlargement of the human type, above all, the contradiction between every moral concept and every scientific concept of life—enough, enough—here is a problem which we fortunately share with not very many persons, living or dead....[10]

In Nietzsche's text, there is a recognition of the shift in moral vision that had taken place in the latter decades of the nineteenth-century. In seeking the limits of knowledge since the Renaissance, pure science had increasingly shut down the measure of the universe in the pursuit of a fully objective, determinate world. But Nietzsche did not have the tools to correct it. Nor did Artaud, four decades later, find the means to do so, in his advocacy of "the plague" to destroy all the hated institutions of the present century:

> If the essential theatre is like the plague, it is not because it is contagious, but because like the plague it is the revelation, the bringing forth, the exteriorization of a depth of latent cruelty by means of which all the perverse possibilities of the mind, whether of an individual or a people, are localized.[11]

Artaud would return civilization to a time of raw, psychological terror: a time of primal nature, inspiring a "contagious delirium" of "immediate gratuitous acts" that have no place in the intellectual content of society, but in which, nevertheless, "all powers of nature are freshly discovered."[12] In the name of this new morality Nietzsche advocated war and evil, so as to effect, in the end, a synthesis of supermen: "Thus the highest evil belongs to the highest goodness."[13] Nietzsche died in 1900—five years before the Theory of Relativity, forty-eight years before the prediction of Complexity Theory, sixty-four years before the discovery of Chaos Theory. He did not live to see the highest evil, World War I, enflame and destroy the old orders of Europe.

ACTIONS, DISPLACEMENTS, AND TRANSFORMATIONS: THE MOVEMENT FACTOR

> There is something in the nature of theatre, which from the very beginning of theatre has always resisted being theatre.
>
> Herbert Blau, *Substance* 37–38 (1983): 143

In his pioneering study of semiotics, Petr Bogatryev declared: "On the stage things that play the part of theatrical signs...acquire special features, qualities and attributes that they do not have in real life."[14] However, Bogatryev's formulation raises as many questions as it provides answers. How can you establish the basis of a theatrical transformation. In the process of identifying attributes and determining relationships between signs and sign objects, what constitutes Bogatryev's "real life?"

In Robert Wilson's *The Forest*, we may speak of the half-animal half-man Enkidu's encounter with the whore, [Geno Lehrer], as "exemplary of primal life force, of brute procreativity, of creation as we know it—at once tender, cruel, sensual and erotic." The implications of the signifiers described are only one set of visual images; there are also music and textual images that must be deconstructed and displaced, then mixed, *cosa mentale*, in the minds of the audience.

As was noted earlier, Wilson's folk operas effect a *displacement* or series of displacements, of audio and visual tracks at a pre-conscious level to set the immediate situation free. The formula for displacement and transformation can be applied to the actor's art. From the perspective of the spectator, this displacement consists of images in the script (the playwright's art) that articulate the language of the play at a sub-verbal level to produce relations that are (or may be, depending on the enthusiasm and participation of the spectator) novel, startling, unconventional, dismal, thought provoking. These, in turn, articulate with the actors' gestures and emotions to produce transformable relations.

At the conscious, intellectual level, a conventional substitution or *transformation* of the artifact, or of the relations of the artifact, produces interpretation, recognition, relevance, social awareness—the semiotic interplay of art and culture, social history, of the aesthetic object in the collective consciousness. Does this constitute reality? In the neural theatricalism of Robert Wilson's *The Forest*, the world behind the text is not endlessly deferred but subverted in an unconstituted and divisible presence. In his stage presentations, he brings multi-faceted potential for images and words to the door of the limbic system, which are then mixed, by neural substitutions and semiotic transformations, and carried through by the spectator, cerebral and limbic processes of an intentional order somehow still intact.

As if to accentuate this posture in aesthetic theory, Herbert Blau's writings in the 1980s stressed the "remainder factor" (as he described it), that lay beyond all conscious formulations. In *Theatre Semiotics*, Carlson summarizes Blau's predicament: "This something, a resistance, a "rub," encourages the never-to-be-realized original experience instead of the re-presentation of performance."[15] Blau's comment that performance gives "visible body to what is not there, not only the disappearance of origin but what never disappeared because it was never constituted," strongly confirmed neural theatricalism, or "neural/post/unconstituted deconstructionism."[16] The artist in performance, "living testimony to a primordial substitution or displacement," as Blau describes him or her, "is born on the side of the Other."[17] From the perspective of complexity studies of evolutionary consciousness, one might conversely argue that Blau's "remainder" is the self-generating origin in the mid-brain! The "other" is the reverse of popular legend: the intellectual processes of the cerebral cortex.

Samuel Beckett describes a similar situation in *The Unnamable*: "Perhaps they have carried me to the threshold of my story, before the door that opens on my story, that would surprise me, if it opens, it will be I, it will be the silence, where I am..."[18] In a similar fashion at the end of *The Unnamable*, Beckett's admonition to his reader: "...I don't know, I'll never know, in the silence you don't know, you must go on, I can't go on, *I'll go on.*"[19] [italics added] unveils a prospective displacement of a series of virtual Others into a reader-constituted presence. Beckett makes the nature of neural intentionality very clear: "The search for the means of to put an end to things, an end to speech, is what enables the discourse to continue."[20] It would seem that movement, inherited from an ancient source, is the end-goal. For the twenty-first century student of evolutionary complexity, *mimesis*, "imitation of an action," has become "displacement of an inner image," a not too subtle update of Aristotle's insightful analysis of dramatic poetry.

THE UNINTENTIONAL ARTIST

Playwrights in recent years have become very vocal about the notion of artistic creation. David Rabe, Neil Simon, Robert Wilson and David Mamet, among others, have all commented in one way or another on the "less conscious" aspects of their artistic work. Their direct witness to the process of writing playscripts that produce intentional thought and emotion confirm recent research on the relationships of artistic creation and spectator comprehension.

In *Speed-the-Plow*, David Mamet's 1988 Broadway block-buster, two Hollywood hustlers—Bobby Gould, new head of production and Doug Brown, an independent producer—manacle morality (and the English language) in the name of the god of movies—*money*, to be made on a prospective script:

Gould: The question, your crass question: how much money could we stand to make...?
Fox: Yes
Gould: I think the operative concept here is "lots and lots..."
Fox: Oh maan...
Gould: But money...
Fox: Yeah...
Gould: Money, Charl...
Fox: Yeah...
Gould: Money is not the important thing.
Fox: No.
Gould: Money is not Gold.
Fox: No.
Gould: What can you do with Money?
Fox: Nothing.

Gould: Nary a goddam thing.
Fox: …I'm gonna be rich.
Gould: "Buy" things with it.
Fox: Where would I *keep* them?
Gould: What would you *do* with them?
Fox: Yeah.
Gould: Take them out and *dust* them, time to time.
Fox: Oh, yeah.
Gould: I piss on money.
Fox: I know that you do. I'll help you.
Gould: *Fuck* money
Fox: Fuck it. Fuck "things" too…
Gould: Uh huh. But don't fuck "people."
Fox: No.
Gould: 'Cause people, Charlie…
Fox: People…yes.
Gould: Are what it's All About.
Fox: I know.
Gould: And it's a People Business.
Fox: That it is.
Gould: It's *full* of fucken' people…
Fox: And we're gonna kick some ass, Bob.
Gould: That we are.
Fox: We're gonna kick the ass of a lot of them fucken' people.
Gould: That's right.[21]

In a 1988 interview, Mamet substantially explained the cynical, quick sand morality of Gould and Fox:

> We live in very selfish times. Any impulse of creation or whimsy or iconoclasm which achieves general notice is immediately co-opted by risk capital, and its popularity—which arose from its generosity and freedom of thought—is made to serve the turn of financial extortion[22]

But elsewhere, Mamet has also said:

> We, as a culture, as a civilization, are at the point where the appropriate, the life-giving, task of the organism is to decay. Nothing will stop it, nothing *can* stop it, for it is the force of life, and the evidence is all around us.[23]

The one statement seems to contradict the other. If Mamet had the first statement in mind when he wrote *Speed-the-Plow*, a director might well interpret the play as a statement of the corruptness and phoniness of the American entrepreneurial spirit in Hollywood, the "get-rich-quick-at-all-costs" syndrome that seemed to pervade the 1980s. But if the director focused on Mamet's second statement, where the life-forces of our culture seemed to be moving towards an inevitable build-up of maximum entropy, in which nothing would halt the pre-determined demise of our civilization, then *Speed-the-Plow* mocked the very notion of creativity. Perhaps Mamet himself was not clear.

In any case, Mamet's commentaries could have led to the production of two different plays. If Mamet's contradictory postscripts are to be believed, the "intentions" of the playwright at the moment of composition are, at best, tenuous threads to follow through the labyrinth of consciousness. In Mamet's case, over a period of time the intentional fallacy dicta would seem to be justified! But even focusing on the play as an ambiguous *objet trouvé* is not sufficient. As one noted director has commented, "some playwrights" *unintentional* fallacies are enough to make one gag."[24] In their *New York Times'* discussion of the role of the unconscious in creating a playscript, Neil Simon and David Rabe stated unequivocally that the role of *conscious intention* in the creative process is extremely limited:

> *Rabe:* I think my conscious mind is not as intelligent as my unconscious. My conscious mind is very much interested in controlling everything and making it orderly—making it orderly in a familiar way. Then the unconscious can come up with something original.

> *Simon:* I know. I know when my unconscious is doing the writing, because when my conscious is doing it, it seems familiar to me when I see it later on…. *Brighton Beach Memoirs* took nine years from the inception of the idea. I let it sit for six years. It just kept going in my mind. I would think about it, and six years later I wrote 35 pages. I said, 'This is good, but I don't know how to write the play.'… And it took another three years. And then I sat down and went right through the play. But the unconscious is doing the work. It's typing away.

> *Rabe: Streamers* took a total of seven years from the beginning. Suddenly I sat down and in about three or four days rewrote the whole play. And it was a full-length play now. I don't know how I knew—there's no way to measure that.

> *Simon:* I feel very happy when I've got an idea for something that I think is worth doing. And then I can leave it alone and not work at all—it can/must do its own work there while I go to the beach or play some tennis.[25]

If Neil Simon and David Rabe are to be believed, the real creative force in playwriting is intentional but not voluntary. Intention is biological as well as interpretational. In the act of creation, an intention, or a series of intentions, must be present. It bears repeating: creative processes are not automatic, beginning from square one for each function, but resemble in remarkable ways the dynamic of the Foucault analogy of relational and associational geological strata. Furthermore, the brain of man is not a computer resembling any of the current generations.

On a more technical level, these elaborate non-limbic secondary association areas, which lead to associational maps of neural populations, are—through the differential amplification of reentry stimuli over a period of time, the basis of human thought as we know it. Parallel processing, the next generation of computers perhaps, is as close as science has yet come to the essential dynamic of imitating human brain function.

Playwrights whose work deliberately explores the mechanisms and possibilities of the bicameral mind pose considerable difficulties for both directors and actors.

As playwright and director, Robert Wilson, for example, methodically creates a visual core on which an audible text is superimposed. Wilson's method is conscious while his creation is less conscious. The playwright describes his creative process as the equivalent of combining the *visual* portion of a silent screen film with the *audio* portion of a radio play.[26] Instead of courting "the unconscious," as Simon and Rabe suggest, Wilson deliberately stalks the unconscious in a profusion of visual images, to which he later appends a text about a central idea. In describing the composition of *the CIVIL warS*, Wilson consciously mapped his deliberate chronology:

> *Wilson:* First I made a structure out of what was earlier arbitrary—some 5 cuts. Then without knowing the text or visual effects, I worked from this diagram. Then I drew diagrams and covered the walls of my empty apartment in NYC with paper and began to make drawings and sketches as set builders would look and collected a series of drawings that filled out the structure. I found out through this the single theme that would thread through [the cuts], until the optical chapters were integrated like a tapestry.... And then separate (sic) from that, I wrote a text without thinking of a definite context.[27]

Wilson is deliberately conscious of his effort to create subliminal audible and optical text. In the foreword to the production playbill for *The Forest*, he elaborated on the problem Newtonian "container-space" poses for contemporary dramatists:

> If we don't illustrate in the audio book the visual imagery of the text, then the visual book of the audio screen is boundless. And if we don't illustrate the visual book with the audio book, then the visual screen is boundless. The problem with most theatre is that it gets boxed in....We're always shifting in and out of these interior/exterior audio-visual screens.[28]

Wilson's efforts to free the audio-visual screens are made to give the spectators a creative voice in the interpretation of the stage performance.

In displacing the playwright's traditional role of providing a definitive script, Wilson forces the viewer to choose a text. By freeing the spectator's reentry creative choices, by making the conscious mappings (and unconscious neural mappings) a personal (and biological) choice of the spectator watching and listening to the stage performance, the playwright provides the audience member with the *opportunity* to have as authentic an experience of the production as the artist onstage.

In *American Voices*, a recent critical assessment of five American playwrights, Esther Harriott suggests that Sam Shepard is himself "a narcissistic individual...[who yearns] to return to a symbiotic union with a dominant parent."[29] Such a statement, made in response to her analysis of Shepard's dramatic works, raises critical alarms. Modern psychoanaytic criticism has shifted its emphases away from the psychology of the playwright to relations between playwright, spectator, text, and language.[30] Inevitably the playwright, when asked, says that he or she "has a spectator in mind," or "sees the play in terms of a spectator looking at the stage."

Positivism has died a slow death in American theatre, whether in the form of Harriott's style of critical analysis; or in the guise of the "benevolent despot" school of directors. It even exists in the dictums of acting commentaries that solely insist on achieving "emotional truth;" and/or in classroom discussions of "schools" of criticism.[31] Radical subjectivity *begins* by asking: "what is the meaning of 'benevolent?' of 'achievement?' of 'truth?' of 'Being?' beyond 'the reality of the doing' in the life of the theatre?" The cognition of the complex mosaic of neural processes is the beginning of artistic judgment.

A less conscious approach to the problem of creativity appears in Esther Harriott's interview with Marsha Norman: "How do you give [your characters] that voice? How do you get inside their consciousness?":

> NORMAN: I'll begin to hear lines of dialogue. And the question gets to be 'Who's speaking?' At that point, it's like that old TV show, 'This Is Your Life'—isn't that the one where they had someone speaking behind the scrim? So at that moment I'm the person sitting there in the chair, thinking, 'I hear this voice, I know that's somebody that I know. Who is it? And then once you get it 'voiced,' once you know how the characters talk, the writing can begin.[32]

Marsha Norman describes what might be acknowledged as Gerald Edelman's "intentional re-entry," where maps of cross-modal association pathways are brought to the primary sensory areas of the brain—neuronal stimuli in search of an object—an act clearly on the mind of Pirandello in his 1925 Preface to *Six Characters in Search of an Author:*

> Creatures of my spirit, these six were already living a life which was their own and not mine any more, a life which was not in my power any more to deny them.... They are detached from me; live on their own; have acquired voice and movement; have by themselves—in this struggle for existence that they have had to wage with me—become dramatic characters, characters that can move and talk on their own initiative.[33]

Cleanth Brooks sees the intentional fallacy as demonstration of the fact—noted by Socrates two millennia ago—that writers sometimes write better than they know, "that often times more than *conscious* design is involved, and that the writer does not always tell...the whole truth...about what his specific intentions were."[34] But he does not seem to recognize the possibility that playwrights like Robert Wilson, Luigi Pirandello, and Samuel Beckett may deliberately court a more—or less than—conscious design.

Samuel Beckett, a playwright who acknowledged privately that his plays were a "less conscious creation, was particularly sensitive to the precision of concrete details that anchor his creations."[35] In *the CIVIL warS*, Robert Wilson insists that a particular piece of stage property be spiked within an eighth of an inch; actors in a Wilson play moved "by-the-numbers" in measured robotic precision.[36]

A 1984 Beckett production at Cambridge's American Repertory Theatre shows how insistent a playwright can be about the appearance of his work. JoAnne Akilaitis's production of *Endgame* elicited Beckett's fierce resistance to the play's gutted Metropolitan Transit Authority underground setting.

Commenting in Paris in September 1985, Beckett remarked to me: "Clov looks out a window; he sees nothing. How can this setting represent that!"[37] Describing the production as "an End-Circus," Beckett refused to acknowledge the production as his work. The relationship of setting to text, in which meaning, related to a particular space, creeps through the interstices of the language, reflected Beckett's absolute sense of the playwright's intentional relationships of space to language. In every case, we have the movement, onstage and in the cerebral cortex, of sound, of sound-images, as they trigger discontinuous neuronal responses in the brain of the spectator. Ultimately these responses are found in the interstices of the brain, deep in the *thalamus* of an ancient brain that preceded the evolution of man and the hominids.

"DANCES OF DARKNESS" FROM ONO KAZUO AND HIJIKATA TATSUMI

The non-western form that most resembles Wilson's imagistic theatre is Butô dance drama, founded in the late 1950s by Ono Kazuo and Hijikata Tatsumi. According to Benito Ortolani, the experimental new dance genre has roots in Hijikata's dadaistic and surrealistic experiments.[38] These Japanese artists have a need to express in a subversive manner the feelings of anguish and terror that each experienced during the wartime destruction of Japan. In performance and in public pronouncements, video tapes of Hijikata's performances confirm this statement.[39]

Ortolani also notes the strong European traits behind the formal training of a Butô dancers: the intensity of training regimes; the surrealistic and hallucinatory atmosphere of unconventional, over-controlled, mostly slow movements of the performers; the near-naked bodies painted white; the shaved heads of the all-male performers; the rolled-upwards eyes; and finally, the wide-open mouths, reminiscent of early Expressionist paintings.[40]

On May 24, 1959, Hijikata Tatsumi's dance piece, *Kinjiki* (Forbidden Colors), was performed in Japan. One commentator, Goda Nario, maintains that "if this performance had not taken place, today's modern dance world would be an entirely different place."[41] Goda describes the performance in the following words, "The dance was performed by two males, a man and a boy; a white chicken was strangled to death over the boy's crotch and then, in darkness there were footsteps, the sound of the boy escaping and the man pursuing him." Goda's commentary continues:

> Although *Forbidden Colours* only lasted a few minutes, it combined the barbarous act of strangling a chicken with the treatment of the anti-social and supposedly taboo topic of homosexuality. It made those of us who watched it to the end shudder, but once the shudder passed through our bodies, it resulted in a refreshing sense of release.[42]

There is no indication that this dance has ever been repeated. Nor should there be—ever! The infamy that Hijikata's performance garnered was apparently countered by the reputation of this artist for serious work, and the realization that the linkage of contemporary social and cultural conditions to the deep structures of the human mind was an unexplored resource. In enacting his sense of Japanese consciousness in the post World War II state, Hijikata deliberately set out to redraw the boundaries of conscious experience.

In other performance pieces, defiant of traditional modesty of Japanese society, Tatsumi Hijikata has depicted semi-nude Japanese women gyrating back and forth, tongues lolling, saliva dripping from the corners of their mouths. Superimposed on this scene are images of Hijikata and his voice-over, in Japanese, with translated English text, haranguing the assembled: "We break rules and upset forms, we suspend decision making, we let the body speak for itself, reveal itself, we reject the superficiality of everyday life."[43]

Butô is a consciousness that angers and repels as much as it convinces. The dance master tears off the veneer of polite society to expose the roots of behaviors; to lay bare the limbic system; to return Japanese society to its earliest pre-history, some 29,000 years ago. In that regard, Butô has much in common with prewar German Expressionism, with 16th century Scottish Nationalism.

In his swan song to the world, Hijikata described his work in the following words: "We shake hands with the Dead, who send us encouragement from beyond our body; this is the unlimited power of Butô." In 1986, as if to punctuate this statement, while performing a dangerous descent, Tatsumi Hijikata fell to his death head first from a rope suspended from the top of a tall building.

Adherents to Butô style dance performances soon emerged. Sankai Juku, which means "studio of mountain and sea," is a second generation company that draws its inspiration from ancient Japanese mysticism. The dancers of Sankai Juku trace their way back through the body to the origins of existence. Gestures border on the extreme, and attempt to open up space to a time before human consciousness. There, in the midst of primeval chaos, something can always be born, appear, live and die in the space of a moment. But what is this consciousness portrayed? And how does it work?

The most spectacular portrayal of pre-history comes in Sankai Juku's production of *Jōmon Shō* (Homage to Prehistory), created by Amagatsu Ushio. In that dance, four males, nearly naked, are suspended from the top of the proscenium by ropes which slowly lower their painted white bodies to the stage. There, they unravel from their fetal positions in barely perceptible slow motion. Amagatsu Ushio's vision of the distant past is also a vision of a collective unconscious, as conceived by the troupe.

If a message is to be found in this piece, it is that there was a unity before the individual self was traumatized by modern existence. As this commentator described it, "the work shows a child traumatized by powers it cannot control, by an adult and an old woman, alienated and isolated and finally swept away by catastrophic change." These images are contrasted to those of communal bliss and ecstatic bliss "to be had when one loses one's sense of individual self in the great cosmic un-consciousness—the bottomless and formless (or, rather preformed) I".[44] The dance troupe toured widely in the United States in the 1990s.

THEATRICAL THRESHOLDS OF NEURAL DARWINISM

As noted earlier in this text, when we do theatre, we replay the history of the species that extends far beyond the emergence of consciousness. When we examine the range of species since the Cambrian Age, five hundred millions years ago, there is the probability of 17 trillion kinds of creatures. Nothing like that number ever in-habited the earth. The record as it stands today, according to Jay Gould, shows too few vertebrates, no invertebrates after the vertebrates emerged, and the odds that, "if we replayed the tape of species on earth, the chances of getting *Homo sapiens*— bipedal, two-eyed, conscious creatures—are practically nil."[45]

However, Gould may not have considered important complexity studies factors in his calculations: man's discontinuous neural evolution, periodic climatic and ga-lactic catastrophes, modifications of Darwin's "survival of the fittest" theory, as they impact Newtonian regularity in classical biology.

We can define "two-eyed," and "bipedal." But when we attempt to define con-sciousness, we run into the key problem. How do the physical events occurring in our brains, while we think and act, relate to our subjective sensations?—that is, how does the brain relate to the mind and to a unitary consciousness, which is what we want to portray onstage in any given moment? In *Neural Darwinism*, Gerald Edel-man makes persuasive arguments for natural selection of neural groups and the creation of "neural maps" in the *cerebral* cortex as the basic synthesis of uniquely in-dividuated man. The essential question Edelman asks is: "How does the variable brain deal with an infinitely complex world—an 'unlabeled world'—to make sense of things?"[46]

More immediately, how does consciousness arise from the mass of neurons? In terms of theatrical consciousness, how does a performance arrive from the disparate parts of consciousness, as we know it? How do we choose the mask against which certain character traits are highlighted, or the setting against which certain events can be dramatized, or the lighting or costumes or music against which certain emo-tions can be portrayed? And finally, how does the spectator synthesize the opposi-tion between mental representations and reality, and "uncover" the performance? It now appears that the *prefrontal lobes* orchestrate the brain's behavior, and from that sense of rhythms arises the notion of consciousness.

Problems of neural mapping, reentry of sequential differentiated impulses, primary and secondary repertoires, cell adhesion molecules that regulate bonding between cells to create variability in neuronal groups, and substrate adhesion mole-cules, as well as associational categorization, pale in comparison to the dynamic equations of brain activities involved in completing an idea, or perceiving a concept. Linear equations cannot manage all the variables.

Any measurement of a system produces chaos-organized randomness. As typical of weather forecasting as it is of brain waves or theatrical performance, whether conscious or not, onstage the actor, the set designer, the playwright all conspire to introduce a certain element of randomness. Ambiguity is a key element, so as to stretch the interpretation of a given moment beyond a spectator's expectations. Therein lies suspense, terror, pity, laughter, joy and relief, where otherwise lies only a human body, standing in a space in front of a lot of other bodies.

If man is seen as a theatrical species, chaos is the very foundations of the theatrical enterprise. In the brain of man, consciousness is evoked by the discontinuous firings of neurons in chaotic responses along mapped neural pathways. The trigger is not in the *cerebral cortex* but in the *thalamus*. Interestingly, there are no special qualities of nerve matter in the brain. This has important consequences for viewing the state of chaos as endemic in our body systems. As Antonio and Hanna Damasio have noted, molecules of the brain are not unique [to the human brain], but are composed of common amino acids. Likewise, "no new principles or molecules specific to the brain have emerged in studies of hormone regulations or of trophic factors that influence the survival and differential of neurons."[47] In summary, once the *cerebral cortex* is complete, and the *pre-frontal lobes* are functional, it only remains for neural connections to every other part of the brain to be finished, and human consciousness can begin

CENOZOIC SONATAS

The weather's fine. Let's go out and play! For the last 90,000 years, weather has been sporadic. For the last 65 million years, with the disappearance of the dinosaurs, more play activity emerges, more species' phyla diverge, more successive temperature challenges occur, and more opportunities arise for *Homo sapiens* to evolve into modern humans with language, speech, theatre consciousness and societal goals. Meteorology is a prime shaper of evolution.

We can track the evolution of ourselves with the emergence of moderate climates; intense global warming 160,000 years ago, followed by declining temperatures 110,000 years ago; then warming again until 90,000 years ago until the height of the last ice age, 17,000 years ago.[48] By 8,000–7,000 years ago, the great glaciers covering the northern hemisphere retreated. A "civilizing" warming trend began about 5,000 years ago, reaching its peak 3,000 years ago, as modern culture took root in hospitable river valleys like the Nile, the Tigris, the Indus and the Yalu.

In an earlier section, I described the connection between the human voice and birdsong. While birds do not possess the sound-image connections that make human speech possible, a recent article in *Science Journal* makes it clear that: "The love of music, that unslakable, unshakable, indescribable desire to sing and rejoice, rattle and roll, is not only a universal feature of the human species, found in every society known to anthropology, but is also…far more ancient than previously suspected."[49] And more widespread among the species of earth! Orcas, for one species, should also qualify. In studies conducted by Dr. Patricia Gray, a researcher at the National Academy of Science's Biomusic program, and Jelle Atema, a researcher at Wood's Hole Marine Biology Lab, flautist, and "reconstructor" of ancient flutes, circa 53,000 years ago, they found definite music scales.

More importantly, these musicians/researchers also found that songs sung by birds and humpback whales "converge on the same acoustic and aesthetic choices and abide by the same laws of song composition as those preferred by human musicians, and human ears, everywhere." Gray notes that male humpback whales vocalize over a range of seven octaves, and yet confine their singing to lilting musical intervals; "in other words, they sing in key:"

> They also follow a favorite device of human songsters, the so-called A-B-A form, in which
> a theme is stated, then elaborated on, and then returned to in slightly modified form.

The haunting effects of Racine or of Molière at the Comedie Française, sung by the actors, is an evolutionary remnant of birdsong or whalesong. One can also look at the speeches of Shakespeare for the A-B-A pattern, where the Stratford bard states a premise, follows with the requisite action, and then concludes with a premise summation: A-B-A in prose or verse, in action as in thought (both *ethos* and *dianoia*). In our own time, Samuel Beckett's plays can be argued on the same basis: dissonance in *Endgame*; resonance in *Happy Days*; assonance in *Waiting for Godot*.

Birds too, have their own composition specifics, using the same notes, rhythmic variations, harmonic patterns and pitch relationships as man. Dr. Gray, who reportedly composes movements "for saxophone, piano and whale," waxes lyrical on the pentatonic scale which is used by the hermit thrush: "the pentatonic scale is the scale on which the prehistoric flutes are built, and it's also the basis for a lot of rock 'n' roll music today."

Gray's commentary suggests emphatically that language and the human brain is not a unique phenomenon, but an evolutionary development from pre-history, tens of millions of years old. The fact that only male vertebrates sing suggests an origin in mating rituals that evolved into general communication, none moreso than *Homo sapiens*, the species with a discontinuous *cerebral cortex* that could integrate sound and images into a thing called human language. The driving force, as I noted previously, is a need to communicate an inner sense of self, summarized as the rhythms of consciousness.

There is, of course, much research left to do. Dr. Mark Tramo, a neuroscientist at Harvard Medical School, and his colleagues have not yet discovered a specific "music center." Magnetic resonance imaging may find the differences between man and birds and sea mammals. Trano argues that neural structures that participate in musical experiences are also players in other forms of cognition, auditory and otherwise. He suggests the *left planum temporale*, critical for perfect pitch, is involved in language processing along with fMRI scans that indicate an interplay between right and left hemispheres of the brain. Whether this is a consequence of the *corpus callosum* connecting both halves, or a divided site for processing, remains undecided. Whatever the final result in the biomusic debate, the evolutionary discontinuous *cortex* and the "executive brain functions" of the *pre-frontal lobes* probably separate human language and human consciousness from all other species on earth.

ON THE BRINK OF INSTABILITY

To the question of why a male mating call from the era of the dinosaurs developed into human speech, there can be no direct answer. Certainly birds and predecessors of *Homo Sapiens* diverged long before 65 million years before the present.

All that is suggested here is the certainty that the core of a power structure that would influence selection by females of dominant males was extant long before the species. The singular pattern would await the maturity of a discontinuous cortex and the emergence of a capacity to integrate sound and sound-images before human speech could emerge. Connectivity-options abound across the species. The miracle that consciousness as a rhythm could emerge as well from pattern systems also led to theatrical consciousness. (The latter system depends on intrinsic redeployment of experience.)

The question that remains is not one of direction but of figure/ground relationships. Evolutionary changes that result in new formulations and even new species cannot be discarded. We think great thoughts and define grand ideas, but our feet still arise from dust and ashes. Consummate actors tred softly and with reverence towards their inner beings and bring back with them the traces of reptilian ancestors that are and were always with us.

One would be hard pressed to find a better analysis of cerebral functions in the brain than that proposed by Scott Kelso.[50] Noting that the human brain is intrinsically chaotic, possessing by definition an infinite number of unstable periodic orbits, Kelso insists (rightly so) that this flexibility in a resting state allows for an infinite variety of responses. In *Dynamic Patterns*, Kelso summarizes his position from the beginning:

> The brain is fundamentally a pattern-forming, self-organized, dynamical system poised on the brink of instability. By operating near instability, the brain is able to switch flexibly and quickly among a large repertoire of spatio-temporal patterns—a "twinkling" system creating and annihilating patterns according to the demands placed upon it.[51]

Kelso expands on this central thesis, derived from C. S. Sherrington's "enchanted loom" analogy of the mind of man as a constantly dissolving pattern maker to

Hermann Haken's theories of pattern formation in non-equilibrium situations, to postulate an extraordinary summary.

The spontaneous dispersal and re-routing of old patterns into new formulations, as the brain seeks momentary equilibrium, seems to fuel the mind. Kelso repeats his adage: "Neural maps of the brain constantly shift their relationships," according to Kelso, "to form new collective activity responsive to external factors."[52]

This kind of neo-Darwinian flexibility in biological evolution allows creatures with relative coordinated systems on the margins of instability to thrive, while those that are hard-wired die out. Theatrical consciousness is a celebration of the taming of 160 million years of evolution, and may well represent the ultimate test of reality. Samuel Beckett reminds us there is freedom in the interstices of consciousness[53]; Henri Poincaré speaks to us of science that uncovers the relations among things; "outside these relations there is no reality knowable."[54] The scientific quest for the nature of our world, whether described as freedom or reality, where the best we can achieve is sensory relations, is the goal of every theatre artist as well.

It would seem that theatrical consciousness is a development of reflection on movement. Movement and considered response is a mediated event in man. Immediate response excludes any form of intention. On hearing a sharp noise, most animals will instinctively react. But where intentions, which arise in the *superior frontal gyrus* of the *cerebral cortex*, play out a role, whether real or staged, *neural activities precede a motor discharge.* This event shows up, according to Kelso, even when subjects are asked to think through a movement but not asked to perform it. Consider the world class athlete mentally preparing to do his thing, or a skier, contemplating a run, or even a golfer, mentally working through the stroke.

This phenomena, I submit, is the basis of *catharsis,* a "thinking through" with the attendant neural activities, of a moment. For a spectator in the theatre, the *supplementary motor area* of the *superior frontal gyrus* contains the reality of the act (whether executed or not). The imitation the actor displays leads to something called the "readiness potential," which begins in the cortex about one second before the response. (Kelso indicates the *cerebellum* and *baslal ganglia* are also active, (which would accord with evolutionary origin principles.)[55] In life, the final motor cortex external discharge takes place; in the theatre, only "Ralph" follows through. The important thing is that the BP: "Bereitschaftspotential" exists as an intentional act whether or not the act is completed. (Kelso, it should be noted, began his life as a professional child actor in Dublin, Ireland; he completed his education in America.)

"Play your intentions!" is an instruction all actors get from their director, and with good reason. Intention is an imitation of movement and of reflection that predates 65 million years, to a point in time when movement was replaced by "thinking through" a movement. The resultant motor cortex discharge without movement is *catharsis* in the theatre. As Aristotle first noted, *mimesis,* "imitation of an action," has both an outer and an inner truth. In theatrical consciousness, living vicariously is possible. Evolutionary movement, as I have discussed above, is always the essential key.

As is well known, we have a choice of interpretations of *catharsis*, following Gerald Else or Butcher: to "purge one's feelings" through experience of such feelings displayed onstage, or to "learn from the experience" of observing such feelings onstage—the expulsion catharsis or the knowing catharsis. Aristotle probably leaned towards a knowing catharsis, since he used that "pleasurable experience that derives from learning" as the basis for challenging Plato's rejection of theatre in his Republic.

Catharsis, whatever its source, clearly involves some kind of thought process derived from external experience, an extraction from what is present to an inner source. Since the *neocortex* was not extant in a species that would not exist for another sixty million years, the speculation becomes impossibly moot. Of necessity, we have to look much deeper into the reptilian brain and related parts of an intentional discontinuous act that would eventually become the human brain, a "little lower than God."

LOOMS' LIMITS

When I look at your heavens, the work of your fingers,
the Moon and the stars that you have established;
What are human beings that you are mindful of them,
Mortals that you care for them?
Yet you have made them a little lower than God,
and crowned them with glory and honor.

(Psalm 8: 3–5)

Scott Kelso's incisive views on the nature of dynamic pattern formation in open non-equilibrium systems is a major step forward from 20th century conceptual diagnosis of the nature of brain and mind.

The debate on computational programs duplicating thinking, and fixed neuronal group selection, supported by Gerald Edelman and colleagues, cannot go forward. Reaching back to the beginning of the last century, the modern interpretations to consciousness proceed, in many ways, from Sherrington's enchanted loom, "where millions of flashing shuttles weave a dissolving pattern always a meaningful pattern though never an abiding one."[56]

One could say with reasonable surety that the century-old debate would not have taken the direction it did had not Nobel prize winner Sherrington, and his disciples, Nobel Laureate Sir John Eccles and Scott Kelso, among many others, not followed the research of an inspired teacher. But the conclusion is improbable.

From whence should patterns arise and fall away? The image-bearing loom is captivating. The solutions to recovering the flow of constantly shifting patterns has been attacked with extreme vigor, in physiological and mathematical modes, and with intuitions borne of decades of experimentation. Those experiments include delayed response insights of Neils Bohr, models of chaotic attractor systems by Russian mathematician L.P. Silnikov, and Roger Penrose's theory of an unknown linkage between quantum mechanics and special relativity. One must insist that complexity studies are a prime component for future investigations.

These correlations are based on the principle that, where there is smoke there is a mirage of fire. Excluding the *prefrontal lobes*, the consciousness mirage is located in the desert of our *cerebral cortex*, deposited over 160 million years of evolutionary process. The reason why relations rise and vanish as we approach is simply because the reptilian core cannot be ignored. Beckett's notion of freedom in the interstices of thought reflects this vintage origin.

Writers and artists have always perceived the majesty of this consummate work of nature. The grand shuttle is a fulsome translation of our evolution, and we will find the rhythms of complexity and evolution intruding on neural patterns from their source, however faintly, as consciousness and discontinuity are allowed to influence the dynamic organization of brain structure.

"Pattern" is of course the correct assumption and conclusion of process in the *cerebral cortex*. But rhythm of the *basal ganglia* environs is the origins of consciousness, of all that is left of the heritage we cannot forget but dare not ignore.

When we look at theatre; when we do theatre, we come as close to delving into our basic civilizeable nature as anything can come in this life. We find there, at one instance, the pain-wracked Antonin Artaud, making his "birdcalls," a little lower than God, but trekking towards his own version of fossil imprints:

> kré
> kré
> pek
> kre
> e
> pte
>   ...   o recho modo
> to edire
> di za
> tau dari/ do padera coco[57]

Part Three

❧❦☙

The Theatrical Genome

All passionate language does of itself become musical—with a finer music than the mere accent; the speech of a man even in zealous anger becomes a chant, a song.

Thomas Carlyle

Chapter Nine

❧❦❧

Ancient and Modern
Theatre Consciousness

SAMUEL BECKETT'S NEURAL COSMOS[1]

Sweat and mirror notwithstanding they might well pass for inanimate but for the left eyes
which at incalculable intervals suddenly open wide and gaze in unblinking exposure long
beyond what is humanly possible.

Samuel Beckett, Imagination Dead Imagine (12–13)

In a 1992 report, entitled "Problems and Prospects for Theatre Research," Joseph Roach discussed the interdisciplinary range of theatre scholarship—from the perspectives of the new historicism, cultural poetics, poststructuralism, hermeneutics, reception-aesthetics, semiotics, feminist studies and cultural studies—in the light of what he calls, or cites as, "the performative act."[2] His commentary was echoed by John Rouse who, in his declaration that "all signifying systems used by theatre and drama are always already part of other cultural texts," offered not a program critique but a basic reiteration of an aesthetic for the application of semiotics to performance studies.[3]

These clarifying statements embody the most elementary social and cultural codes as foundations on which to build the difficult notions of difference, of signification, and of a subjectivity (that continually emerges as a process and not a product). But they exclude the most important code of all—the bio-communicative code, the intentional nature of which underlies all intellectual and performative behaviors.

One might say that deconstruction came as a warning sign, a Cassandra-like prescience to pilots of vehicular postmodernism, that contemporary figure/ground relations of positivists or postpositivists were not adequate to detail the literary cosmos. For better or for worse, the success of deconstruction as a methodology is directly attributable to the failure of aestheticians to include bio-codes in their social and cultural analyses of texts. Like the "dark matter," recently found in interstellar space, that scientists believe makes up 90 to 97 percent of the universe, and will allow the universe to stabilize at some point, bio-communicative matter, and codes for its behavior, is the missing ingredient in performance studies.[4]

How to incorporate that code into socially and culturally coded studies is a question of some magnitude. That scarcely noted biological codes exist in literary works, underpinning social contracts and cultural hegemonies across the literary spectrum, is singularly demonstrated in several works by Samuel Beckett, Robert Wilson, and many performance artists—Joseph Beuys, Laurie Anderson, Ping Chong, et al., and works—*Theme for a Major Hit*, *Primary Demonstration: Horizontal-Vertical, Opposing Mirrors and Video Monitors on Time Delay, Going Thru the Motions*, and *Einstein on the Beach*, to name a few of the first generation of "body-artists".[5]

One cannot truly begin to assess social and cultural codes on which postmodern literary theory is based until the bio-communicative codes are acknowledged and addressed. To do otherwise would be to lead us back into the postmodernist dilemma, to deconstruction and the disassembling of normative comprehension as end products in themselves. In the interests of opening the envelope on this topic, a short analyses of bio-codes in Beckett's *Endgame* follows.

One might well ask, "What does biology, in particular the 'New Biology' have to do with social and cultural codes in the Theatre?" There are a number of answers. In the first place, the latter two codes are applicable from many humanistic perspectives. Lawrence Stone, Princeton Professor Emeritus of History, has given one answer. He recently intoned that "every cultural enterprise, even science, is at least in part a social construction."[6] In the sciences, Harvard biologist Jay Gould, vehemently agreed. "Science," said Gould, "is done by individuals whose conclusions are influenced by the beliefs they bring with them."[7] In "twice behaved" behaviors, avatars of the performative act prescribed social and cultural codes as an article of faith as the necessary step in deciphering the new interdisciplinary studies. But how can we make the leap from biology directly to theatre? The answer comes in part from Erwin Schrodinger, the renowned quantum physicist, who in the 1940s almost single-handedly brought to biology experimental truths that may take us to the edge of the universe.

In Ireland in February 1943, under the auspices of The Institute at Trinity College, Dublin, Erwin Schrodinger delivered the quantum biology lectures that extended his elegant 1926 theory of quantum physics into the field of molecular biology. He described the unfolding of quantum events not as the certainties of Newtonian mechanics, but as an undulating wave of possibilities. In 1944, Cambridge University Press published an edition of these lectures, entitled *What is Life? The Physical Aspect of the Living Cell.*

In 1943, Samuel Beckett was in hiding from the Gestapo somewhere in the south of France, possibly at Roussilon in the Vaucluse. Where Beckett obtained the text is unknown, but he regarded it highly. As it turned out, Schrodinger's small text began a revolution in more than molecular biology. In 1992, Professor L. C. Lewontin of Harvard described this particular text as "the ideological manifesto of the new biology."[8] In conversation in 1985, Mr. Beckett acknowledged with a wry grin the importance of Shrodinger's small oracle to his own work.[9]

Criticized by Irish friends like Jack Yeats for being amoral in his writings, Samuel Beckett, in his refusal to cater to the modernist, absurdist, existential tides of his generation was, on the contrary, profoundly engaged in transforming Schrodinger's bio-codes into a theatrical framework. His affinity for pictorial artwork, and his friendship with painters and sculptors—Jack Yeats, Bram and Geer Van Velde, Avigdor Arikha, and Henry Hayden in particular—was matched by his enthusiasm for quantum science.

In June, 1946, Beckett obtained a copy of Schrodinger's quantum biology lectures, and gave it to his uncle, Dr. Gerald Beckett.[10] Examining the deep structures of Beckett's works through Schrodinger's quantum biology lens clarifies a number of puzzling details. *Imagination Dead Imagine, How It Is, Endgame,* and *Play* are only a few of Beckett's works that are anchored in the bio-artistic framework that begins with *Waiting for Godot.*

The question might then be asked, "How *does* the new biology relate to theatre, specifically, to Beckett's theatre?" One broad response is that theatre is indigenous to the human species. It is not that Beckett was unique in his uses of biological codes: all playwrights and all good actors have, for all time, consciously and unconsciously, found their sources in bio-codes. But these sources are seldom recognized in critical writings, or even acknowledged in production.

Writing arose as a means of exchanging goods and fostering trade. But the capacities to speak and write about themselves arose and persisted for the broad benefit of the species. As Antonio and Hanna Damasio describe it, "humans and species before them had become adept at generalizing and categorizing actions and at creating and categorizing mental representations of objects, events and relations."[11] As can no other species, we can plan the future; we can judge the past; we can contemplate the present. The question of learning from the past is quite another matter, as Unamuno once suggested.

Far from being pre-ordained, even in a biological sense, we are a product of random evolutionary circumstances that Harvard biologist Jay Gould has called, in an evolutionary context, "punctuated equilibrium." Five heads, each with a different function, for example, or three limbs, to propel us forward in circular motions (perhaps another version of Beckett's "headlong tardigrade"), could have been viable physiological options. As Gould explains: "The contingency of evolution does not depend on the random nature of genetic mutation. It arises because mutations have qualitatively different effects, and because these effects can be amplified…."[12] Only the remarkable intelligence of a tiny creature—a predecessor to *Homo sapiens*—on a primeval forest floor marked the potential for our species' survival.

The display of this intelligence quotient is a part of any "twice-performed" performative or behaved act that Roach noted in his brief. Even more than the intelligence factor, when we do theatre we are replaying a biological heritage that extends well back into more recent pre-history, some two and a half million years ago with the evolution of the first direct descendents of modern man. Even more germane, the performative act onstage celebrates the development of "fine tuning" in the *pre-frontal lobes* in the last eighty thousand years. That history, and the residues of that past, let me repeat for emphasis, are the necessary ingredient for reinventing or re-stating or reexperiencing the moment. I speak of movement on the performance stage, or within every cultural and social act on the world stage.

In the last year of his life, anthropologist Victor Turner came to the realization of the crucial importance of neurobiology as a basis of performance studies. He did not have an opportunity to augment his critical research into the cultural dynamics of performance with MacLean's "triune brain" thesis, for example, or other studies that, in fact, had been developing independently for some years. In the arts them-selves, the research had been in full swing for decades.

The *Homo sapiens* story began in the Cambrian Age, some five hundred and seventy million years before the present with the appearance of shelled, carapace animals. Those punctuated mutations evolved into modern man's precursors some two and a half million years ago. Looking again at the immediate and long past his-tory of the earth, John Wheeler, quantum physicist at Princeton suggested, "The deepest lesson of quantum mechanics may be that reality is defined by the questions we put to it."[13] This comes as no surprise to playwrights and performers who have long and similarly framed the dramatic experience. The actor's key is to find in per-formance the residue, the remnants of experience from a long, geological past time, and to initiate or reinvent his or her recall of those same particles that impelled the original moment.

In this context, the actor's art is not reducible to scientific analyses of intellec-tual constructs, or even of emotional recall of particular conscious scenes from the near or distant past. The residues lie infinitely deeper. If the actor finds a voice and a performative act for the residue, and the spectator is drawn infinitely within, the performance may be transcendent—and magical. It may not be considered objec-tively analyzable. In the case of the *limbic* system of man, there are four centers of focus (and no more): fight, flight, sexual encounter and hunger stimulus. There is no question that theatre, working out from those four centers and expressing the needs and conditions of human life, is any less indebted than our species to the track of photons as they approached a "galactic beam splitter" some fifteen and a half billion years ago. Waiting to be subjected to experimental forays, conducted by unborn beings on a still nonexistent planet, the neurobiology necessary to reflect on that state of being has only been available to the human species in the last milli-second of a twenty-four hour solar clock.

The emergence of *Homo sapiens* across the evolutionary track of species was unique and fundamentally unpredictable. When one examines evolution, it is immediately apparent that the figure of an actor appearing onstage in a performance arena is only one tiny fragment of an epic story, the likelihood of which is manifestly unreasonable. The citation from Samuel Beckett's *Imagination Dead Imagine* indicates the deep "Schrodingerian" structure in the playwright's own creative works, and his profound interest in amplifying segments of quantum biology in an artistic venue.

In looking at the mechanisms of genetic development, "on the average, only the 50th or 60th descendent of the egg that I was," Schrodinger set the deep past into present-day contexts. The physicist remarked on the marvels of "the visible and manifest nature of the individual, which is reproduced without appreciable change for generations, permanent within centuries—though not within tens of thousands of years." Carried forward by the species and within the species at a temperature that has not varied from the beginning, Beckett's text begins (with an explanation): "No trace anywhere of life, you say, pah, no difficulty there, imagination not dead yet, yes, dead, good, imagination dead imagine"(7).

In Beckett's narrative, something takes place beneath the level of intellectual conception, *something* in a place as deep as Schrodinger's pseudo-microscopic analysis of the DNA molecule. Beckett's processes of temperature and of light are Schrodingerian: "Combining in countless rhythms, [they] commonly attend the passage from white and heat to black and cold, and vice versa…until, in the space of some twenty seconds, pitch black is reached and at the same time say freezing-point. Same remark for the reverse movement, towards heat and whiteness."[14] Even Beckett's pendulum swing of temperatures is Schrodingerian: a return to stasis, "rediscovered miraculously after what absence in perfect voids," against the background of "the *little fabric*" (emphasis added).[15]

In *What is Life?* with a nod to quantum mechanics, Schrodinger suggests not linear continuity in the species, but genetic molecular *discontinuity*. At a given moment across the span of many centuries, there is not a gradual evolution, or survival of the fittest, but everything *suddenly changes*. This sudden change is also the great surprise of Jay Gould's punctuated equilibrium theory.

For Schrodinger, "the great revelation of quantum theory" was that features of discreteness were discovered in the Book of Nature that defied the continuity anticipated by classical physics. The transition from one of these configurations to another was not a smooth transition but a quantum leap. Schrodinger explained: "If the second one [had] the greater energy (a higher level), the system [had to be] supplied from outside with at least the difference of the two energies to make the difference possible."

In *Imagination Dead Imagine*, Beckett, following Schrodinger's example of discontinuous biology, obliged with his own bio-artistic imagery: "Piercing pale blue the effect is striking, in the beginning. Never the two gazes together except once, when the beginning of one overlapped the end of the other" (13).

Beckett's contribution to the revolution was to envision this overlapping span in a concrete, bio-artistic frame "from the fraction of the second to what would have seemed, in other times, other places, an eternity"(9). How did Beckett acknowledge this leap? Returning to the figures, each lying within his and her semicircle, he described the circumstances of a multi-generational quantum leap: "Sweat and mirror notwithstanding they might well pass for inanimate but for the left eyes which at incalculable intervals suddenly open wide and gaze in unblinking exposure long beyond what is humanly possible" (12–13).

There is some evidence to suggest that, at a dramaturgical level, Beckett wanted the intentional energy, and the skeletal frame of interpretation, to come from the outside, in the spectator's arena. That energy was, as might have been predicted, the eye of the observer: the "unique inquisitor" in *Play*, "E" in *Film*, Bom or Pim in *How It Is*, and a range of observers from Vladimir in act two of *Waiting for Godot*—"At me too someone is looking...."—to *Imagination Dead Imagine*—"...and at the same instant for the eye of prey the infinitesimal shudder instantaneously suppressed" (14). On the level of the "new biology," Beckett explored the possible discontinuous movements in the gene pool of the species: "...to see if they still lie still in the stress of that storm, or of a worse storm, or in the black dark for good, or the great whiteness unchanging..." (14).

As Beckett so expressly indicated, one need only use one's imagination—not the dead imagination of classical physics but the living imagination of the Book of Nature that Erwin Schrodinger deciphered, and Beckett so eloquently described. By adding an adjective and a noun phrase to the opening lines of *Imagination Dead Imagine*, the meaning becomes clear: with "[Classical] Imagination Dead, Imagine [the Book of Life and quantum genetics...from a discontinuous perspective.]" (I hasten to add my trepidation at joining the legions of commentators who have added text to Beckett's, where none was ever necessary!)

The last phrase of *Imagination Dead Imagine*: "...and if not what they are doing," returns the inquiry to the Schrodinger text, and the wait for the next eon of time and the contingencies of evolution. In another ten thousand years, qualitatively different mutations may develop in the human species, born of quantum leaps in the diploid molecules of Samuel Beckett's imaginative formulations—a discontinuous change, perhaps once in ten millennium.[16] In *Imagination Dead Imagine*, the revolutionary picture of DNA, of the new biology, of Schrodinger's quantum formulations of the living cell, is described in detail (along with Beckett's deep concept of our discontinuous future), decades before its general acceptance.

A second thesis, and possibly of more interest to Beckett (and to us), is Schrodinger's adaptation of positive and negative entropy—the basic behavioral principles of the living organism—to quantum physics and to what has now become the new molecular biology.[17] Described by Schrodinger as the characteristic feature of life, the state of maximum entropy is a dynamic equilibrium: "Every process, event, happening—call it what you will: in a word, everything that is going on in Nature means an increase of the entropy of the part of the world where it is going on."[18] To avoid death, according to Schrodinger, a creature feeds upon negative entropy, drawing sustenance from the environment. In so doing, every living organism succeeds in freeing itself from all the entropy it cannot help producing while alive. A living organism continually increases its entropy—or, as one could say, produces positive entropy—"and thereby tends to approach the dangerous state of maximum entropy, which is death."[19]

Living on the edge, while not fraught with the same degree of risk for us as for creatures of primeval epochs, is still with us, generally glossed and sugar-coated for the most part with cultural and social codes of behavior—if you don't eat, you die. And if, in the rise of ethnic tensions the socio/cultural codes are stripped away, the bio-communicative codes remain, in the shattered fabric of nations in geographies all over the globe, as we have so recently seen.[20] (It might be suggested as a hypothesis, that our socio/cultural codes are the "fine tuning" by the cerebral cortex of basic instinctual limbic responses that underlie all actions. Given an opportunity for limbic responses, in a world that has dislodged political and social order, we reach back for another form of stability—to the basic instinctual survival patterns that kept and abided our ancestors, at the expense of rival clans, generational differences, or unfamiliar ethnic backgrounds.)

While refusing to engage in political or social or cultural consequences of the breakdown of codes, Beckett put the biological code principle to work in a typically striking visual way. The most dramatic propositions of the unavoidable positive entropy of life, old folk in various states of decay, litter the landscape of Beckettian dramaticules.

Two examples illustrate a range of Beckett's thinking on this topic: in *Happy Days*, Winnie's mound is an expression of the build-up of positive entropy (in act one, up to her waist; in act two, up to her neck.) Clov's first speech in *Endgame* has a similar origin: Zeno's pile, "Grain upon grain, one by one, and one day, suddenly, there's a heap, a little heap, the impossible heap."[21] (In *Happy Days*, the reference to Zeno's pile is only the more apparent basis of Beckett's thought processes.) Put rather simply, the meaning of life as an expression of desire is set against the inevitable working of the biological principles that create life in the first instance. As a basic formulation, the apparent contradiction of death growing stronger in the midst of life is not a solipsistic statement but the very foundation of Beckett's theatre.

A third issue of importance, directly related to Schrodinger's influence on Beckett, came as a post-scripted chapter to the Trinity lectures' text. In his "Addendum on Free Will," Schrodinger describes the self as "not I" but "the canvas" of life that we interpret as self.[22] The "canvas" opens a world of possibilities for Beckett, culminating in the 1972 premiere production of his play, *Not I*, at New York City's Lincoln Center.

For Schrodinger, our notions of self—of "I"— are not a substance at all, but an interpretation of experience that is promoted to the level of "I" at any given moment. So where is the permanent trace of life? In our minds only, or in a succession of minds, or in the indeterminancy of being that we historically regard as substance: "Not I—me!" is Mouth's third person summary of this situation in *Not I*. The new biology, and possibly the developing dramatic theory of chaos, insists on the analysis of this basic hermeneutic as a keystone of man's relations to his environment. The conventional deconstruction of self, the unavoidable subjectivity that Derrida posits as the political consequence of socio/cultural agendas, has no meaning at the bio-communicative coded level.

To confront bio-codes is to render postmodernist theoretical structures that are based on the familiar socio/cultural codes, as secondary influences of signification, at best. This is particularly true in Beckett's theatre, in which stages of an image evolve before the spectator's eyes, (a methodology that Beckett developed from the writings and paintings of Jack Yeats). It is also true in the dreamscapes of Robert Wilson, the biological horizons of Laurie Anderson—set to music [I would really enjoy her responses to Rabbinical questions, put to her at intermissions!], or to Ping Chong, or the Blue Man Group. In the state of theatre as lived life, one cannot, finally, separate Dreamkeepers from their Dreamsongs in the Australian outback (for example) that traces Song lines of an Aboriginal walkabout to 40,000 years before the present.[23]

As Clifford Geertz so astutely notes, the "genre blurring" of conventional nineteenth-century disciplines in the postmodern era must be replaced by performance categories that refigure our socio/cultural agendas.[24] Moreover, this performance agenda demands, in my opinion, that critical theorists reach beyond the paradigms of Cartesian mind-body dualities, and even beyond the paradigm of Husserl's *Logical Investigations* that effectively deconstructed both the "epistemological subject" and the "objective world."[25]

The anthropological basis of much of cultural performance critique is foregrounded by Geertz, and the late Victor Turner. Great attention has been placed, and a great academic industry begot, on Turner's pronouncements in the study of the cultural dynamics of performance. However, it is also true that this inspired anthropologist, embracing at the end of his life the biological consequences of man's triune brain, showed us a glimpse (sadly, too late for him to develop its implications) of the bio-neurological and bio-communicative codes under discussion. Were Turner still with us, undoubtedly he would have updated the new biology concepts that play a critical role in the breadth of theatre "performative studies" today. Perhaps not in a study of Samuel Beckett and his works for the theatre, but certainly Turner would have continued his anthropological investigations of the triune brain that first emerged in 1970.

Beckett was apparently intrigued with DNA biology, as described by Schrodinger in 1943, and subsequently labeled as "the grail of molecular biology."[26] Decades later, Harvard Professor Lewontin updated the missing pieces of a forty-nine-year-old debate, begun at Trinity College, by noting that DNA is a dead molecule: "It is also not self-reproducing, and it makes nothing. And finally, organisms are not determined by it."[27] What it does do is serve as a template. Lewontin explains:

> Reproduction of DNA is, ironically, an uncoupling of the material strands, followed by a building up of new complimentary strands on each of the parental strings…. The role of DNA is that it bears information that is read by the cell machinery in the production process. Subtly, DNA as information bearer is transmogrified successively into DNA as blueprint, as plan, as master plan, as master molecule.[28]

Beckett stunningly anticipated this entire sequence in *How It Is.*

Positing the replicating double helix of the DNA molecule in many specifics, the invariable sequence of *How It Is* works in a simplified eight-part series (four-part victim, four-part tormentor). The speaker/narrator (1) flees Bom, (2) approaches Pim, (3) unites with Pim, (4) torments Pim, (5) is fled from by Pim, (6) is approached by Bem, (7) is united with Bem, (8) is tormented by Bem, and beginning again, flees Bem.

Lewontin described this process: "By turning genes on and off in different parts of the developing organism at different times, the DNA creates 'the living being' Body and Mind."[29] In Beckett's "bio-dramatic grail," Pim and Bom know each other only by reputation: Pim as the victim, and Bom as the tormentor, of the narrator. They never meet, but serve in their sequential uncouplings as "blueprint"-victim, and as "master molecule"-tormentor to the narrator. United in the interests of torment, separated in the event of suffering, these DNA couples are relegated in Beckett's unrelenting bio-dramas to present the spectacle of life's most basic confrontations.

In every case, the tormentor bears information of the persona of a prior tormentor, which he inflicts upon a new victim, to "transmogrify" this ex-victim-turned-tormentor at the next stage. There is not time to give more than an outline of the process. Speaking in the third person, in a text without punctuation, "before Pim," the speaker/narrator carries a prior identity into part two, awaiting instructions for "life the other above in the light said to have been mine…no one asking that of me never there a few images on and off" (8).

A series of images begins with, presumably, a young Samuel Beckett saying his prayers at his mother's knee: "I steal a look at her lips/ she stops her eyes burn down on me again I cast up mine in haste and repeat awry".[30] This is followed by an image of a girl and a dog on a grassy mound: "the girl too whom I hold who holds me by the hand the arse I have...again about turn introrse fleeting face to face transfer of things swinging of arms silent relishing of sea and isles".[31] The last glimpse is of a youth "pale staring hair red pudding face with pimples protruding belly gaping fly spindle legs sagging knocking at the knees wide astraddle for greater stability feet splayed one hundred and thirty degrees fatuous half-smile".[32] These images from scenes in the light above mix with an immediate reality of a tongue lolling in the mud, of the sack "where saving your reverence I have all the suffering of all the ages",[33] and filled with sustenance—Pim's "tins" of "miraculous sardines."

The narrator's means of locomotion by crawling in the mud, "right leg left right arm push pull",[34] continues until—at last, with the left hand "clawing for the take instead of the familiar slime an arse two cries one mute".[35] Pim has been overtaken. The former tormentor becomes victim; the narrator—as former victim—the new tormentor. One could scarcely ask for more concrete and detailed description of the breath of life, or of the genetics of continuity of the species. At the level of bio-communicative codes, there is no need for intellectual folderol. At the same time, there is also no need to limit or to exclude interpretations of the action by social and cultural codes, (and the deconstruction of those same images to demonstrate the illusion of truth, unity, origins and closure of any situations).

Bio-communicative codes function as permeable membranes at the boundary of language and of image formation. They appear in the guise of intentional relations, whose intra-subjectivity is multiple and shifting. Bio-codes are not reducible because they are not constructed, but are always a part of what has already been in the act of becoming; a part of the intentional act that shapes as it is itself transformed—the "Not I" of "me".[36]

In part two of Beckett's *How It Is*, the narrator replaces Pim's memories with his own templates of life, "namely the canvas upon which [experiences] are collected." In developing the analogy, part three in *How It Is*—Pim's departure—corresponds to Schrodinger's "youth that was I, [whom] you may come to speak of...in the third person." The sequence of how it was before, with, and after Pim's arrival and departure can be interpreted as Beckett's artistic development of Schrodinger's metaphor for the evolving self: "you may come to a distant country, lose sight of all your friends, may all but forget them; you acquire new friends,...you still recollect the old one that used to be 'I' but is now no loss at all."[37]

Part three of *How It Is* is life after Pim, the escaped former victim, and before the arrival of Bom, the speaker/narrator's next tormentor, with the voice of us all "without quaqua...alone in the dark the mud end at last of part two"(99). A moment of reflection occurs before the sequence renews itself. Bem reaches for the narrator, to begin part three, as the narrator reached for Pim in part one: "Bem had come to cleave to me see later Pim and me I had come to cleave to Pim the same thing except that me Pim Bem me Bem left me south". With Bem Bom's arrival the narrator acquires Bem's name, and a borrowed life "said to have been mine above in the light", [38] moving in procession like some medieval Chaucerian *tableau vivant;* a vast train of beings, where:

> at the same instant I leave Bem another leaves Pim and let us be at that instant one hundred thousand strong then fifty thousand departures fifty thousand abandoned no sun no earth nothing turning the same instant always everywhere [39]

and thereby living out the instructions of the bio-code of the last ten thousand years, from a residue of one hundred and ninety-five million years.

"Outside time and space without extension," Beckett's characters have become quantum molecules in one dimension, failed habits of life in a more immediate sense, but inevitably depicted in real terms, in real life that transcends epistemological abstractions or Newtonian mathematical equations.

Reproducing without appreciable change for generations—centuries if not for ten thousand years, ("and borne," as Schrodinger notes, "at each transmission by the material structure of the nuclei of the two cells, which unite to form the fertilized egg cell,") the biology of *Homo sapiens* is exposed at its quantum bases.[40] The intentional, hermeneutical marvel of that premise is the basis of Schrodinger's *What is Life?* and of Beckett's *How it Is.* DNA is a template passed to the next generation, and expressed by Beckett in the familiar social and cultural codes of the theatre, with its deeper, permeable, discontinuous, core code intact.

Schrodinger's "little canvas" of "experience and memory" is also extensively dramatized in *Endgame.*[41] One of the key images of the play, Hamm's bloodied handkerchief, frames the action. From the opening dialogue, we are aware of a very significant prop:

> (...Pause. Hamm stirs. He yawns under the handkerchief. He removes the handkerchief from his face. Very red face. Black glasses).
> HAMM: ME—(he yawns)—toplay.... (2)

The stage directions and last lines of text of the play complete the "handkerchief frame":

> ...speak no more.
> (He holds handkerchief spread out before him).
> Old stancher!
> *(Pause).*
> You...remain.
> *(Pause. He covers his face with handkerchief. Lowers his arms to armrests, remains motionless).(Brief tableau).Curtain* (84)

In this strange play about "zeros" in perspective, of creatures dying of darkness, the use of a bloodied handkerchief to conceal Hamm's face is, on the surface, equally enigmatic. There are no running sores, no cankers or ulcers, as in earlier pieces. There is no external bleeding, no hint of a physical calamity to "staunch." The simple explanation for this is that the damage is metaphysical, every bit as painful, and

inflicted by social and cultural codes. Beckett explains the human dilemma with a
short vignette from his principal player in this drama. Hamm tells the story of a
mentally disturbed engraver who, having seen the rising corn fields and the sails of
a herring fleet, now sees only ashes. According to Hamm, he was "spared."

Although Hamm offers no explanation of exactly what is "spared," one explanation might be that the presumptive world of hope, brought on by a vision of transcendent beauty, and socially accepted expressions of that beauty, no longer has the
power to seduce the intentions of a more primal code.

Momentary salvation comes only in the recognition of "the game"—an endgame to be sure—where one can freely recognize the necessity of weeping "for
nothing, so as not to laugh, and little by little...you begin to grieve."(68) The lessons of bio-communicative codes are harsh. Occasionally Hamm falters, as in the
moment when he asks Clov for certain words from the heart:

> Stop, raise your head and look at all that beauty...that order.... Come now, you're not a
> brute beast, think upon these things and you'll see how all becomes clear.... (80)

To which Clov (significantly addressing not Hamm but the spectators) replies:
"what skilled attention they get, all these dying of their wounds." (81)

Those wounds, of course, are the welts and flayings of socio/cultural codes,
masquerading as the core of significations when in fact they are infinitely transformable markers of supreme indeterminancy. We all die of earnestness, Clov
seems to say, and ignore the reality that lies beneath conscious intellectual awareness. Wounds of involuntary memory, which Beckett learned as a young man when
he published his short work on Proust,[42] cover Hamm's face, bloodying him all over,
as surely as they bloodied Pim in *How It Is*. To go on with life, Hamm stanches the
wounds with his handkerchief, a manifestation of Beckett's deepest cry, of an earth
"extinguished though humankind never saw it lit"(81), and of Beckett's consolatory
benediction to the living from the dead:

> One day you'll say to yourself, I'm tired, I'll sit down, and you'll go and sit down. Then
> you'll say, I'm hungry, I'll get up and get something to eat. But you won't get up. You'll say, I
> shouldn't have sat down, but since I have I'll sit on a little longer, then I'll get up and get
> something to eat. But you won't get up and you won't get anything to eat.
> *(Pause)*.
> You'll look at the wall a while, then you'll say, I'll close my eyes, perhaps after I have a little
> sleep, I'll feel better, and you'll close them. And when you open them again there'll be no
> wall any more. (36)[43]

All of these speeches combine in the central image of identity, of self, of relations to others, of memory and of experiences of "I" as a knowable bio-coupling.
The maddening collection of discontinuous significations of recognition (in neurobiological terms, the set of neural maps), is precisely Shrodinger's neutral "canvas,"
upon which the data is collected. After long years of self-determination, Hamm's

pathetic discovery that, "what you really mean by 'I' is merely that ground-stuff upon which [experience and memory] are collected."[44] Hamm's agony of recognition that past lives are no longer germane, even while "hamming it up on-stage," or on the stage that Beckett takes as exemplary of life itself, is the last phase of life's processes.

The bloody handkerchief signifies the pain of giving over to other selves and facing the humiliation of amounting only to "that ground-stuff," a canvas of possibilities that successively fade and, over eons of time, are successively discounted. There are no options. Hamm's multiple images create fractal zones of non-linear series of successive "I's," of DNA molecules outside of our immediate time-space continuum: "Infinite emptiness will be all around you, and all the resurrected dead of all the ages wouldn't fill it, and there you'll be like a little bit of grit in the middle of the steppe."(36).

The "little bit of grit" also relates to Zeno's paradoxical "impossible pile." From the vantage point of the spectator, Beckett's text in *Endgame* marks the beginning in theatre of complexity theory, in operation as a self-contained, discontinuous act.

In terms of chaos theory, *Endgame* demonstrates the quantum biology of self, indicated by the *fractal* boundaries of communication of the self, as an integral part of the processes that can best be described as "discontinuous neural mappings of the brain of *Homo sapiens*," with consummate artistic skill.[45] The dissipative images added up to a lifetime of successive "I's." The discrete differences at each stage of the self have created not a depth of personality but a Mandelbrot pattern. With the addition of the handkerchief, "bloodied all over," an expression of Hamm's internal predicament in staving off the final build-up of positive entropy, the thrust of this memory play is clear. In a universe of chaos, the play of intellectual codes that relate and relay our unintentional memories is all we have.

Composed primarily of discontinuous endgames, cultural codes fill our days, bruising us as they careen through our psyches. What better form of play do we have than theatre play? In Schrodinger's *What Is Life?* the declaration of the canvas of the self, the exploration of positive and negative entropy is the characteristic feature of life. The first manifesto of quantum biology changed the course of Beckett's literary career.

It has taken us five hundred years to understand the geography of earth; it may take us fifty more years to understand the evolutionary neural geography of pre-civilization. Now, with the COBE satellite and new space explorations by the $145 million Microwave Anisotropy Probe mission—dubbed MAP for short—of the primordial structure of the universe, we may come to know (extraterrestrially and intraterrestrially as well), the evolution of seven and a half billion years on earth.[46] We will also begin to understand the evolution of the universe, both from the perspective of where we came from and where we are heading, through twenty-six dimensions of folded space-time.

The discontinuous self-generating biological code would seem to be the mother of all codes. We are all descendent of the ancient *Pikea*, a chordate gelatinous worm from the Cambrian Age.[47] As William Calvin so eloquently quipped, in his thesis on human brain development: "winter is key to brain enlargement in 2.5 million years.[48] Winter once a year, an abrupt climate change every few millennia, and an Ice Age every hundred thousand years will speed things up—if you've got our kind of brain."[49] But we sing our song, we—as Beckett's Pim and Krapp sing their songs—must look into the deep structures of the brain and further into the primordial past.

The thought is universal: Bruce Chatwin describes the Australian Aboriginal *Dreamtime* in his own words:

> The Ancients sang their way all over the world, They sang the rivers and ranges, salt-pans and sand dunes. They hunted, ate, made love, danced, killed: wherever their tracks led they left a trail of music.[50]

Trying to make sense of life and understand a world of inexplicable consciousness, Beckett's Pim seeks an explanation:

> a means of noting a care for us the wish to note the curiosity to understand an ear to hear even ill these scraps of an antique rigmarole[51]

Remarkably, Beckett also gives us his answer, as complete as any that science has yet devised. If we have the wit to temporalize the succession of discontinuous moments that constitute our self-generating mind, poised always on the brink of instability:

> in reality we are one and all from the unthinkable first to the no less unthinkable last glued together in a vast imbrication of flesh without breach or fissure (139)

The principles of self-organization in nonlinear dynamical systems, as Scott Kelso insistently reminds us, cannot be forgotten—first or last.[52] Beckett's "antique rigmarole" is more testimony that music is the original form of speech, and that "everything spoken is accompanied by an inner song," a song that has its origins in our midbrain and brainstem 195 million years ago.

DANCE THEATRE FORMS FROM THE INDUS VALLEY

A central thesis of this text is that rhythm has been at the core of theatre from the beginning. Consciousness and theatre consciousness are deep rhythms that derive from the primordial brain as a consequence of the detection of movement, and plans to attack or to thwart an attack. Sexual encounter and hunger are variations. By 120,000 yBP, the primordial brain had traversed the continents, bringing with man the common ancestry of consciousness, whose only variation was cultural adornment. As Sherrington updated might say, "The great cosmic shuttle was weaving threnodies from the beginning."

The first expression of theatre in the Indus Valley came in dances, addressed to and conjoined with the gods. Indu Shakkhar, citing commentary in A.K. Coomaraswami, *The Dance of Siva*, suggests:

> Dancing came into being at the beginning of all things and was brought to light together with Eros, that ancient one, for we see this primitive dancing clearly set forth in the chord dance of the constellations and in the planets and fixed stars, their interchanging, intervening and orderly harmony.[53]

According to Shakkhar, relics of this dance have been found in excavations at Mohanjoddaro and Harappa some five thousand years ago (3,000 B.C.). As a hypererotic form, the art was practiced in the name of the gods of India much as the Maenad Greeks danced in the ecstasies of Dionysus, the god of wine. According to legend, Siva danced for creation and destruction, Visnu, in the garb of the danseuse Mohini, seduced Siva, and Krisna, darling of the milkmaids, danced with all of them in rapture!

Shakhar claimed there was a Negroid race, c. 3500 B.C., that arrived in the Indus valley and spread over North India, only to fall victim to Aryan hordes.[54] A prosperous tradition of settlements, with established dancing traditions and a matriarchal system, these "Dravidians" to the south, and "pre-Aryans" to the north, developed dance forms that might be termed the forerunner to Sanskrit drama.

Early Vedic religious practices included initiates acting out the roles of the gods as cosmic dancers, in which the world was created. This might explain the basis of religion and ritual theatre, but there was evidence to suggest that, through the use of epic recitations, the latent possibilities of drama were evoked.[55] In parallel to the Etruscan dancers who brought mime and rope dancing to Rome in 360 B.C., there are reports that, as early as the fourth century B.C., pantomime came into being on the Indian sub-continent. Finally, by 140 B.C., all the elements of drama were in place, involving a combination of a tradition of epic recitations, dramatic conventions permitted by a Brahmanic culture, and stories of the Krsna legend: "in which a young god strives against and overcomes enemies."[56] The art forms of performance had reached parallel states of development in Western and Non-Western cultures. But there, the similarities ended. Custom, culture and social demands on the peoples demanded very different performance modes.

Western theatre extolled for a time the actor as interpreter of community values. In Greek theatre in particular, the relations of man to the gods—as agent, as subject to, as preyed upon victim—is a demarcation of the works of Aeschylus, Sophocles and Euripides. In contrast, Brahman culture censored entirely the dance-drama traditions. A strict religious and esoteric doctrine was imposed on the drama. Advocating a life of denial, this life was regarded as a mere passage to the next world. As a consequence, and due to the indifference and class prejudice of Brahmanic civilization, drama developed an artificial character, "drifting away from reality," as Skakhar described it, "and looking more and more to court patronage for survival."

As a consequence, there are no surviving physical theatre spaces from the earliest years. If western prehistoric and classical theatre practice is any guide, temporary structures or natural amphitheatre, or cave theatres for primitive rites may have been the norm, far from any authorities' critical preying eyes. But there is no evidence that a popular, thematic theatre emerged, as it did in western form in Dionysiac legends. Varadpanda claims that, although the art of building was understood and practiced between 3,000 and 4,000 B.C., when Mohenjodaro edifices were erected, no remnants of playhouses remain.[57] In fact, the earliest evidence of formal theatres came only from the second century B.C., with simple four-pillored wooden structures, used as dancing halls. Rather than elaborate panoply, theatrical representations consisted of a young woman dancing on the stage accompanied by a four-piece musical orchestra.[58]

To this day, the link between theatre and mythology is still a part of the living culture of India, particularly of rural India. Never developing a truly secular presence, drama was, and remains—as it is described in the *Natyasastra*— a *yajnya*, a "religious ceremony."[59]

Reflecting this religious basis, even at its peak, Sanskrit drama never focuses on the realities of an external conscious world. Unlike Aristotle's western model for theatre as *mimesis*, "imitation of an action," the content of India's three great epics, the *Mahabharata*, the *Ramayana*, and the *Puranas*, suggest the ideal as an "imitation of a state of being that transcends life." The ethereal consequence is that production is geared to an educated elite, who are themselves trained to appreciate the finer points of performance. A language of gesture develops, without which the subtler emotions can not be translated.

Through the arousal of these emotions, a state of eternal bliss is attained, via the triple agency of *Dharma, Artha,* and *Kama.*[60] The goal of this elaborate panoply of forms is the attainment of *Rasa,* defined by Shakhar as a pleasurable state or mood in the minds of the readers and spectators aroused by recitals or dramatic representations.[61] Through the many functions of an acting tradition that includes gesticulation, movement, expression, costume and emotional demonstrations, a highly structured form of theatre evolves in India quite unlike any other form of drama. The precision of the actor's movements and speech delivery are guaranteed to eliminate the expressive, highly erotic dances that date back to the pre-dawn of Aryan beginnings in the fourth century B.C.

At the same time, in an exceptional manner, this elite form of drama also manages, or suggests a means of managing, to develop by the tenth century A.D. a complete identification between the actor and the spectator—a very modern notion. All the other baser feelings are clarified and the heart is metaphorically attuned (as Shakhar expresses it) "to the object of liquid or a transcendental state." By any other name, the *catharsis* of Aristotelean drama is also the prize sought after by Indian dramatists and actors in performance. The major difference in the respective dramas is the high-toned specifics of the actor's "liquid" art in Indian drama for a select, highly cultured audience, versus the expressive freedom of the Western actor for a typically broader audience. Both dramatic forms aim for achievements in culture that many civilizations have sought, few have achieved.

THE CHINESE PEAR GARDEN

In addition to the emergence of culture on the Babylonian plain, in the Nile flood plain and valley, and in the Indus Valley, the most unique and interesting ancient development of theatre was in the Yalu River Valley of North China. Unique because of its self-imposed isolation from the West, Chou Hsiang-Kuang notes that ancestors of the Chinese race lived in a territory south of the Caspian Sea and migrated eastward in the twenty-third century B.C.[62] Interesting, because we have in this culture the first formal evidence of written theatre. With a likelihood that this ancient civilization was an offshoot of the original Sumerian civilization, whose origins are in pre-Babylonian sites on the Euphrates plain, modern Chinese civilization begins in Central Asia in the basin of the Yellow River sometime in the third century B. C.

Jacques Garnet's survey of ancient China notes the existence of theatre dance forms as early as the eighteenth century B.C. through the eighth century B.C. Dance forms and animal sacrifice predominate. According to Garnet, "Traces exist of ancient dances which seem to have been peculiar to brotherhoods of cattle-breeders and there are inscriptions which very often mention sacrifices of several dozen sheep and oxen."[63]

In close agreement with Hsiang-Kuang, Garnet notes that ancestors of *Homo sapiens* lived on the eastern slopes of Asia from the Mesolithic Period (25,000 B.C.) to the Neolithic Period (10,000 yBP). The period coincides with the retreat of the present Ice Age from the middle latitudes, and a time of marked climatic changes for these newly fertile regions. In Garnet's view, the first agriculture and stock breeding occurs in the fourth century B.C., in the wooded valleys of North China and the Yellow River basin.

This was a hunter-fisher population. To the south lay a more backward people, perhaps *Homo neandertalensis* stock in South China, belonging to the Paleolithic period.[64] Not able to compete with *Homo sapiens*, or to endure the isolation and privation that forces the adaptive mechanisms of modern man, Neandertal man disappears in Europe. Whether described as victim of another species' punctuated evolution or as a by-product of a saltatory evolutionary leap, the South China stock also disappears. By the fourth century B.C., a nomadic civilization is in evidence. They live in temporary dwellings, create red pottery, produce millet, barley and rice, store grains in earthen jars, raise pigs, dogs, some sheep and oxen for food, and decorate pots with geometric motifs—similar to motifs found in Neolithic cultures of the Ukraine and Turkmenistan.[65] Do they have language and make theatre?

In this regard, "pot decoration" might be regarded as a forerunner of written language. Certainly symbols of ownership, evidence perhaps of commerce and of private possession, mark a transition point in Chinese civilization. Chang notes that symbolic marks appear on pottery during the early Yang-Shao culture (5000–3000 B.C.). Displays of shaman-like creatures wearing snakes, with dragons on their ears and hands, or "a fish design" dangling from each ear of a human face, are examples of a prelude to written language, and certainly an indication of quasi-dramatic shamanistic rites with religious significance.[66] Sumerian language begins with the impress of trade exchange problems, and only later becomes the basis of culture and the arts. There is ample precedent to believe the same experience greeted China, coupled to the shamanistic rites and courtly dances, which only later became the basis of drama.

A recent discovery of a new civilization in ruins at Annau, a site near the border with Iran and only eight miles from the Turkmenistan capital, Ashgabat, shows that these people had writing or proto-writing, around 2300 B.C. The importance of this archeological dig is clear. Victor Mair, from the University of Pennsylvania, puts the discovery into context:

> You can say we have discovered a new ancient civilization. At the same time, the pyramids of Egypt had been standing for three centuries, power in the Tigris and Euphrates Valley was shifting from Sumer to Babylon and the Chinese had yet to develop writing.[67]

The belief is that this lost civilization was an early trading station on the route of the famous Silk Road as it evolved from the second century B.C. to the sixteenth century A. D. The Annau ruins are more evidence that language began as an instrument of trade, and later spread to general culture.

The actual origins of Chinese writing and of literature are in dispute. Hsiang-Kuang notes that traditional history begins in China with the Emperor Fu Hsi (2852–2738 B.C.), credited as the inventor of some written forms.[68] However, K. C. Chang suggests the Hsi period is in the mythical past. The invention of writing, described by Chang as the recording of events with symbolism, took place sometime before 5000 B.C. But it was not until the myths and legends of ancient China were recorded in the Chou dynasty (1100–256 B.C.), was there anything like a systematized record of the past.[69] A. C. Scott suggests a more recent date, with records of the actual substance of the past extant only since the T'ang dynasty (A.D. 618–906).[70]

Garnet counters with the suggestion that civilization in China began only in the Bronze Age, 1800–600 B.C.[71] An earlier commentator, Walter Gorn Old, agreeing with Hsiang-Kuang, suggested that the historical record began in 2943

B.C., with the Shu King historical record. This date inaugurated the second patriarchal dynasty and the reign of Foh-hi, successor to the mythical reigns of Yaou and Shun immediately after the deluge.[72]

With these multiple perspectives, the prospect of separating reports from a mist-shrouded past, from recorded contemporary events, in order to lay the groundwork for dealing with various authorities' declarations of origins is dim, indeed.

Ancient China was a recognizable civilization two thousand years before the Christian era. Three dynasties played a significant role in creating the Chinese culture: the Hsia dynasty, 2200–1750 B.C., the Shang dynasty, 1750–1100 B.C., and the Chou dynasty, 1100–256 B.C.[73] In reading texts on this subject, it is noteworthy that, while all dates are approximate, only the dates for the ending of the Chou dynasty correspond in every detail. This suggests that recorded history *really began* around 1100 B.C. Prior to the Hsi dynasty (2200 B.C.), myths concerning the foundation of the Chinese nation note the Three Sovereigns: the first was Fu Hsi: "The First Man;" the second was Sui Jan: "The Inventor of Fire;" and the third was Shen Nung: "The Inventor of Plant-Husbandry." Five Emperors follow the Three Sovereigns. The first two are of particular interest: Huang Ti: The Yellow Emperor, "Initiator of Civilization;" and Chuang Hsu: the "Emperor in whose hands heaven separated from earth."

In this mythological enterprise, the origins of being, of civilization, of patriarchy, of cooking and of providing for the group, of civic and religious responsibility for the people, are introduced. We can propose that Sui Jan, the "fire inventor," is related to archaeological evidence of the continual usage of fire in China, from circa 400,000 years ago, beginning at a moment of transition from *Homo erectus* to *Homo sapiens*, or possibly carried as an oral tradition of *H. erectus* down to the prehistoric period circa 30,000 years before the present. To date there is no concrete evidence available. As noted above, lessons for life and for theology are implied, but data on applicable events in the history of real time are completely absent.

The more interesting materials concern the emergence of shamanism, itself a remarkable form of theatre, as early as 2500 B.C. Divination is a popular concern of Chinese emperors, and is repeatedly used for certain ritual and political purposes.[74]

The question-posers come in several guises. Professor Chang notes: "The king himself sometimes brings the question to the diviner, but more often he asks the question through a mediating 'inquirer.'"[75] After divination, an inscription is made on the bones. Music and dancing are included in these oracular rites. How they become a part of the culture beyond the emperor's palace is not clear. However, during the Shang Dynasty (1750–1100 B. C.), Chang reports that during divination ceremonies, animal offerings are made accompanied by strong drink, "to help in the heaven-earth crossings."[76]

This dance "drama consciousness" as William Dolby characterizes it, begins as possible adaptations from other locations.[77] Crossing geographical boundaries like the South China Sea, home of the Ma'yong of the ancient kingdom of Ligor in Malaysia, or the Ural Mountains, home to dance drama from ancient Turkey (which may itself have derived from rituals of the Uralo-Altaic Shamans) or, prior to sixth century B.C., the Aegean of Western culture, and the familiar Dionysian rites of Maenads on the south slope of the Acropolis near Athens, drama consciousness flourishes in distinctive forms. By the time of the Zou [Chou] Dynasty (circa 1027–256 B.C.), shamanistic practices and court dances merge. Court sorcerers invoke the gods, as Dolby describes it, "by means of highly erotic songs" that are enacted dramas in their own right, complete with costume and make-up. In a similar vein, court dances are regularized adaptations of orgiastic religious dances, composed of set movements "that sometimes involved the enactment of dramatic scenes." By the end of the Zou Dynasty, a decentralized government, corrupt and licentious in itself, could not control the warlords who now set up their own court establishments, with their own "court rituals and music" and their own court jesters and wise fools.

The situation parallels the existence of improvisational players, like the Etruscan dancers in Rome in the fourth century B.C. or the performers of Attelan farces, the *fabula Atellana* of Campania who entertain the Roman legions down through the 1st century B.C., or the commedia dell'arte troupes that flourished in 16th- and 17th-century France. Drawing on his particularly narrow definition of theatre, Dolby still finds no solid evidence for real plays.[78] The first sketch of drama, according to Dolby, comes during the Han Dynasty (206 B.C.–A.D. 220) with a playlette known as *Mr. Hurdy of the Eastern Ocean*, the story of a young man able to charm snakes.[79] But as he aged and turned to strong drink his magical powers failed. The story became one of the first popular plays in the Chinese repertoire. Taking guidance from the emperor, or from the warlords that control a region, popular tradition in China does not develop until a secular tradition can be maintained. The notion of satire as a basic tool of drama—as it evolved in Greek theatre—was never tolerated in the conventional actor and acting troupe.

Puppets are a noted exception. In the guise of a puppet, the puppet-master can express a certain freedom of speech that is impossible for the ordinary player. The practice of using puppets as a less offensive means of commenting on worldly events leads to an almost predictable crisis during the reign of Emperor Mu (1001–947 B.C.) when a puppeteer, Yan Shih, constructs puppets so life-like that he incenses the emperor "because these wooden figures ogle the ladies at court."[80] The emperor would have put Shi to death until the puppeteer persuaded the emperor the puppets were only sticks of wood. More ominously, during the time of Confucius, some court dwarfs put on a play (c. 832 B.C.) that satirized the then-chief justice so noticeably that he advanced upon the stage and ordered the instant execution of the performers.

Puppets are a mainstay of the Tang Dynasty Emperor Xuanzong, the "Resplendent Emperor" (712–756 B.C.), who had the Imperial Academy of Music devote its energies to producing puppet shows. But it is Emperor Minghuang's establishment of the Pear Orchard Conservatory (A.D. 744,) devoted to the training of musicians and singers for the theatre, that lays the foundations for Chinese theatre training. During this period many types of performance emerge, from tight-rope walking to pole climbing, to horse riding through an alley of sharp knives.[81] During this time, Buddhist sutras are translated into Chinese to encourage the proselytization of the new religious sects. Buddhist preachers use dramatic skills to enliven their sermons, much as 10th Century parishioners, following the mass, ask French Bishop Amalarius to repeat his performance of the passion of Jesus Christ during the Easter trope. In China, following the practices of the Buddhist evangelists and within a century of the Western practices, non-sectarian Chinese also begin to develop drama as a genre, particularly after the incursion of the Mongols into China in A.D. 1234.

With the old Chinese government in disarray, scholars who devoted their careers to public service found themselves unemployed. This educated elite develops a dramatic repertoire both as a means of livelihood and as a subversive means of undermining the Mongol invaders. In a manner reminiscent of *The Triumph of Horus* (circa 1284 B.C.)—an Egyptian play performed in the Temple at Edfu which undermined the imposed rule of Cleopatra, the last of the Ptolemy line—the Chinese scholars' dramas give their countrymen hope.[82] According to Dolby, this new repertoire is a major cause of the early downfall of the Mongol regime, and the basis of a permanent theatre in China.

Chapter Ten

❧❦❧

Twenty-six Strings of Consciousness

BECKETT'S GENOMICS

...there he is then at last that not one of us there we are then at last who listens to himself and who when he lends his ear to our murmur does no more than lend it to a story of his own devising ill-inspired ill-told and so ancient so forgotten at each telling that ours may seem faithful that we murmur to the mud to him

Samuel Beckett, *How It Is* (139)

Drama consciousness in the modern era retraces the work of contemporary biological theories, beginning with the Verticalists' Manifesto, signed by Hans Arp, Eugene Jolas, and Samuel Beckett, among others. A sense of depth led to a need to comprehend the new world of post-relativity scientific origins. In the case of artists, moving to theories of the unconscious of Jung and Freud also led to explorations in pure and theoretical science.

In the case of Samuel Beckett, the new science interests led him to investigate the perspectives of the great physicist and father of the new biology, Erwin Schrodinger's *What Is Life?* Already outlined in some detail, Beckett's artistic expression of "The Physical Aspect of the Living Cell" is both visually and comprehensively graphic. That *How It Is* came public in a narrative form, without a companion theatrical form, or that it waited to germinate in Sam Beckett's mind some twenty years after the lectures were delivered at Trinity, will be told at some point. Beckett's *How It Is* [1] thesis is a cry of rage and of pain that the individuality we all seek is a fantasy.

Divided into three parts, how is was before Pim, with Pim and after Pim, the narrator speaks of each character as successive victim and tormentor of the other— Bem, Bom, or Bem-Bom, as the story unfolds. The sequence is viewed as an orbit, divided in half with two points established at the greatest distance from each other: A and B, comprising AB and BA cords. Each point is a Schrodinger cell (and what we would now call a human DNA genome strip of approximately one inch in length.), Every generation of 20 to 40 years a new union takes place, comprised (in Beckett's dramatization) as successive predator and prey. How does an artist dramatize the human genome: along the AB and BA chords, the points at furthest distance from each other are the blueprints of evolution and since each form also has a victim and tormentor dimension, a total of four continuum forms comprise one cycle. Finally, since each continuum form passes though a stage of victim and tormentor, there must be eight stages in the cycle to ensure "justice" for all.

> victim of number 4 at A en route along AB tormentor of number 2 at B abandoned again but this time at B victim again of number 4 but this time at B en route again but this time along BA tormentor of number 2 again but this time at A and finally abandoned again at A and all set to begin again (118)

It is understood that 4 is the greatest number possible at any time, standing for six million, six trillion, more, or less. The only necessity is that 1 and 2 are in numerical sequence; an infinite number may exist between 2 and 4. The eight steps are described in the text in the following terms:

> as for each of us then if only four of us before the initial situation can be restored two abandons two journeys four couplings of which two on the left or north tormenting always the same in my case number 2 and two on the right or south tormented always by the same number 4 (118)

The description of movement relates to the eight steps as follows: "two abandons" (steps 1,5), "four couplings" (steps 2, 4. 6. 8), and "two journeys" (steps 3, 7). Victim and tormentor succeed one another in linear fashion, traversing in deasil movement—from East to West, following the path of the sun, and an ancient Celtic tradition.

That this "deasil" movement consists of endless pain, of cries extracted and of suffering endured that does not vary nor cease from their beginnings in the ooze of primordial muck, is Beckett's embellishment. Images, conceived of as a celebration of the gifts of imagination, passed on to millions of other beings: descriptions of life in the sun, blue and white of sky, "said to have been mine," joys of young love on a hill-top, of a puppy straining on a leash, are an ode to creation, a sonata to conscious existence, that end in pain and desolation. Beckett's line is clearly not Darwinian but Artaudian. But he found, as no artist in the Twentieth-century did, the deep roots of theatrical consciousness.

For the history of the earth we are little more than a replicating gene machine. Having once planted our seed, we are consigned to live out our lives, to play theatrical games, and to preserve the next generation of gene bearers. What Beckett did not know is the role that proteins play in that struggle and the influence that climate or environment plays in the evolutionary game. Craig Ventner notes this issue as perhaps the most surprising development in the unraveling of the human genome: "We now know that the environment, acting on biological stages, may be as important in making us what we are as the genetic code."[2]

Lest we pass this discussion of climate off as an aberration out of the past that can have no bearing on our own future, consider some recent evidence produced by Dr. Henry Weiss at Yale University. Climatic changes ended the great Mayan civilization in Mexico and parts of Central America in the ninth century; dramatic drops in temperature and a sudden drought led to the demise of the hunting and gathering Natufian communities of southwest Asia-Middle East in the millennium between 12,500 and 11.500 years ago. Examining core samples taken from lake-sediment beds, Weiss discovered that even the Old Kingdom civilization of Egypt quickly vanished, circa 2290 B. C., due to a sudden drought and drop in temperature. He concludes:

> There is mounting evidence that many cases of societal collapse were associated with changes in climate. These climatic events were abrupt, involving new conditions that were unfamiliar to the inhabitants of the time, and persisted for decades.[3]

Consider a nuclear winter that was a natural happening, lasting for several decades, and we can imagine the consequences on a global scale.

Extending this sentiment, the infinite variation in the replication that gives us "human" qualities of reflection, speech, written language, altruism, compassion, cruelty, could vanish in an environmental disaster lasting only a few decades. The fact that we are the recipients of a discontinuous cortex that emerged across the last billion years is a guarantee that such a course could again change directions. When one considers that plants have only 4,000 genes fewer than man (26,000 to 30,000 for *Homo sapiens*) a splurge in protein manufacture in plant formulations could make up the difference in survival on this planet in some impossible future eon.

To end on a hopeful note, genomic scientists have concurred there is surprisingly little change in the past 50 million years in the human lineage. Theatre as a complex adaptive system of human consciousness, is alive and well. But every hundred million years, there is this planet-splitting galactic rock....

PANDORA'S PROTEOMICS

On Sunday, 11 February, 2001, 24 hours before the official announcement, British newspapers broke the story: the human genome project is complete. Pandora's Box is open. The history of the human species and all antecedents are complete. Or so it appeared.

Francis Collins, Head of the National Human Genome Research Institute in Bethesda, and Craig Ventner, President and Chief Scientific Officer of Celera Genomics, in Rockville, Maryland, took credit for solving the last great puzzle of human life. But it has not been a clean sweep. Two caveats are extraordinary. First, the findings: Mankind has far fewer genes than suspected, perhaps as few as 30,000, whereas the predicted number was 80,000 to over 100,000. In term of other creatures and fauna, plants have 26,000, nematode worms have 20,000, 13,000 maximum for a fruit-fly, and 6,000 genes for everyday yeast. Looking at evolutionary change, it would appear that most variations arise as mutations on the

Y chromosome. Males. it would appear, are primarily responsible for the disasters and triumphs of the species.

Equally surprising, there are only a few hundred genes in the human genome that are not in the mouse genome. Not since Montaigne's 17th century essay on the frailty of man has the ego of homo sapiens been so effectively assailed. And there's more: a tiny segment of DNA, perhaps less than one inch of the 6 foot-long strand of DNA that lies wrapped inside every cell of the body, is the full measure of mankind. Of that paltry inch of human genome, consisting of 23 pairs of chromosomes, the vast space of every DNA strand is empty or swarming with foreign bodies, bits of genetic debris, and ancient genes that infected human predecessors millions and even hundreds of millions of years ago.

Described by Rick Weiss, Washington Post reporter, as "a dynamic and vibrant ecosystem of its own, reminiscent of the thriving world of tiny Whos" that Dr. Seuss gave to us as children, there are great advantages to this genetic evolution. Weiss explains: "By comparing the human genome to the genomes of simpler organism such as the fly and the worm, scientists are also seeing with unprecedented detail how just a few genetic innovations helped launch early vertebrates ahead of the biological pack hundreds of millions of years ago."[4]

This capacity has been enhanced by certain pieces of chromosomes that have broken off or degenerated and subsequently attached themselves, with their parasites, to other chromosomes. By tracking these routes, described graphically by Weiss as "penguins on an ice floe," researchers are able to track large-scale genetic changes that correlate with key advances in human evolution, evolutions that have been passed down for a billion years. As Robert Waterston, Director of Washington University's Genome Center in St. Louis, recently said, "We are basically, it would seem, just a transient vehicle for making more DNA."[5]

Examining core samples taken from lake-sediment beds, Weiss discovered that even the Old Kingdom civilization of Egypt quickly vanished, circa 2290 B. C., due to a sudden drought and drop in temperature. He concludes: "There is mounting evidence that many cases of societal collapse were associated with changes in climate. These climatic events were abrupt, involving new conditions that were unfamiliar to the inhabitants of the time, and persisted for decades." In *How It Is* (1963), Sam Beckett said as much: Pim's cries are "a means of noting, a care for us" in unbounded exchanges of "sardine tins."

How is this possible? It's all in the proteins, manufactured by our genes in the hundreds of thousands. Therein lies the discontinuous, self-generating genius of a post-hominid species. The field of *proteomics*, or the study of identity and interactions of human gene proteins, has an even more daunting challenge than solving the mysteries of the relatively simple 30,000 genomes, composed of 3.1 billion base pairs of A's, C's, T's, and G's. To understand all the proteins, it may yet take another decade, although probably not 500 years. Not with the resources at our disposal today—unless, of course, there are more Pandora's boxes within—like a Russian babushka doll.

The $145 million Microwave Anistropy Probe Mission (MAP) was launched by NASA from Cape Canaveral on June 23, 2001 to explore the crackling sounds of our background radiation, sounds of the infant universe at age 400,000 years. For whatever reason we are witness to those origins; more specifically, we are conscious of our witness to the pre-galactic genesis. For the first time, we can begin to see glimpses of where we came from and where we are going, from the big bang to the big crunch, in the staccato glow of back ground radiation, sound artifacts we can track to some 400,000 years after the birth of our fourteen billion-year-old universe. This "hum" also ties us to how we became "human" some 4.6 million years ago, and how we developed theatrical consciousness some 30,000 years ago.

Our consciousness of where we came from in our ascent as man on earth now includes our awareness that we sit on the edge of the universe as well. After fourteen billion years of inhaling and expanding, will the exhale be as long. Are we half-way home? Or have we only begun to breathe.

Our best estimates now conclude that the first stars formed about ten billion years ago. Our sun formed after several generations of heavy stars sped through their life cycles, flinging the basic building blocks back into space, where some of these atoms formed into an interstellar cloud resembling the Orion Nebulae.[6] As for our presence in a knowable time and space, the MAP explorer will measure the background radiation of the universe when it was only 400,000 years old. The afterglow of the Big Bang should reveal tiny fluctuations before any galaxies were formed, fluctuations of temperature that allowed stars to form, our solar system to evolve, and the earth to become a planet circling about our sun. In specific terms, Martin Rees has outlined a scenario "where about 4.5 billion years ago, a new star condensed, surrounded by a dusty disc of gas, to become our Solar System.[7]

There are now limits to our portion of the universe and to our galaxy. In five billion years our Sun will expand to encompass Mercury and Mars and then die, along with the desiccated earth, and the other planets. At about the same time, the Andromeda galaxy which is already falling towards us, will crash into our Milky Way. Much of this material fell into place once the identity of the microwave fluctuations were identified. Stephen Hawking called these microwave fluctuations the revelation of the 20th century. The parallel is not lost between background radiation temperature fluctuations that allow stars to form, solar systems to emerge, and temperature changes in our biosphere that allow the human species to neurologically evolve.

What is truly amazing is the fact that a species, which began life 65 million years ago as a tiny mammal, cowering under leaves and in the underbrush of a dinosaur-dominated world, surviving by the power of a reflective brain, has become modern man, with a reflective brain that can con the post-Copernican universe.

"Life and the earth," I concluded in the 1970s, "is an aberration in the universe," with absolutely no rationale or justification. But listen to cosmologists in the 21st century, speaking of invisible dark matter that comprises 85% of the universe, of the requisite fine tuning of the Big Bang to allow for galaxies to later form, and of black holes, billions and billions of which were created in the first milliseconds of the Big Bang, and be amazed:

> We are used to the post-Copernican idea that we don't occupy a special place in the cosmos, but we must now abandon 'particle chauvinism' as well. The atoms that comprise our bodies and that make all visible stars and galaxies, are merely trace-constituents of a universe whose large-scale structure is controlled by some quite different (and invisible) substance.... We must envisage our cosmic habitat as a dark place, made mainly of quite unknown material.[8]

We are creatures of the universe in a passing phase of complexity theory. Cosmic background radiation is movement of an infant universe, our movement, galactic scale, where this text began, at the Big Bang, before the improbable—and now, the unlikely Big Crunch. Although the pull of gravity throughout space is a major force, there is an unknown cosmic repulsion (the cosmic antigravity—lambda) that seems to be increasing the expansion of the universe. We will go on forever, in a hundred billion years, even though our solar system, our Milky Way galaxy, the Andromeda galaxy and billions of other galaxies will have joined up in a single system, and the original gases of the Big Bang will have been tied up in the dead remnants of stars—black holes, cold neutron stars, or white dwarfs. On the micro- and macro-scale, movement is the key, has been the key, for ten and a half billion years.

IN THE PRESENCE OF NEGATIVE SPACE

As important as any specifics of these events, we are also now theatrically conscious as well. Measurements of the red-shift of light from space tells us that light is speeding up and the universe is expanding. This is refreshing, for it allows theorists to begin to confirm string theory beyond the third and fourth dimension. This "folded space" can have ten or up to twenty-six dimensions without violating string theory. As to consciousness in that unspace, our emerging theatre of probity has already embraced it. As one commentator has noted:

> Many more dimensions of time and space could lie buried at the quantum level, outside our normal experience, only having an impact on the microscopic world of elementary particles. The fantastic aspect to string theory ... is that it not only explains the nature of quantum particles but it also explains space-time as well.[9]

At the core of life is movement, which is where this book began, as theatre. When we look to the origins of theatre we find rhythms in the *protoreptilian* brain, perhaps 200 million years old; when we look at the origins of the universe, we find "vibrations," according to string theory, small loops, about 100 billion billion times smaller than the proton, are vibrating below the subatomic level and each mode of vibration represents a distinct resonance which corresponds to a particular particle. Thus, if we could magnify a quantum particle we would see a tiny vibrating string or loop.[10]

This is a microcosmic and macrocosmic view of our world and our universe. Theatre (and movement) is the measure of civilization: of where we are, came from, and are ultimately going, within the confines of a swarm of particles that comprise 85% of cosmic space that we cannot see and certainly do not understand.

Beyond our cosmic realm, there is evidence that the Big Bang sponsored trillions of black holes ($10-12^{th}$ power) no larger than a single atom, with a mass the size of a mountain. Martin Rees speculates that these putative black holes could spawn and "inflate into a new (possibly infinite) space-time disconnected from ours,"[11] giving rise to a trillion new universes, with a whole new, inconceivable, pattern of physical laws and evolutionary changes, a duodecillion multiverse for starters. What is not included in this debate is a postulate from complexity theory, whereby, once set in motion, an event-horizon has no single point, but can be simulated as a pattern that emerges only after millions or billions of simulations; a rhythm in fact, or a vibration—in string theory—that can never be pinpointed in Newtonian mathematics. In a similar fashion, our evolutionary biology would not have emerged without the expansionary capacity of a succession of adaptable species. At the apex of those species is modern man, whose discontinuous cortex fostered the emergence of human speech and theatrical consciousness.

On the grand scale, the evidence is mounting that there is a grand design for evolution of our species on earth, and for the evolution of galaxies and the stars in our universe. The fine tuning that allowed only certain rates of expansion after the Big Bang, the asymmetry of our universe, where one thousandth of a percent change in the cosmological constant (omega) would have made the creation of stars impossible, brought light to the billions of galaxies. The excess production of protons over antiprotons, and the creation of helium and hydrogen, allows our universe to have atoms, stars and galaxies instead of only massive radiation and dark matter.[12] Irrefutably now, a guiding hand that shaped this intergalactic genesis is also the Guiding Hand that brought us to consciousness in our present state.

Notes

PREFACE

1. Aitchison, Jean, *The Seeds of Speech: Language Origin and Evolution.* (Cambridge: Cambridge U. Press), 4.

CHAPTER ONE: THE DISCONTINUOUS PROCESS

1. Rosenfield, Israel, "A New Vision of Vision," *New York Review of Books*, Vol XLVII, Number 14, September 21, 2000, 61.

2. Rosenfield, "Vision," 64.

3. Paul Davies, in John Brockman, *The Third Culture: Beyond the Scientific Revolution* (New York: Simon and Schuster, 1995.

4. "Post-Doctoral Fellows," *Santa Fe Institute Bulletin*, Vol. 15, Number 2, Fall, 2000, 30–31.

5. Wood, Bernard, cited in Wilford, John Noble, "Skulls Found in Africa and in Europe Challenge Theories of Human Origins, *The New York Times* on the web, August 6, 2002, 1, of 4.

6. As for the choice of male versus female for the first upright creature, human genome studies now reveal that males are responsible for 95% of the innovative changes in the species. Females thrive in a steady state, while males seem to thrive when innovations are possible or necessary for the species' survival.

7. I use "MYbp" to indicate a shorthand form "million years before the present" throughout this text.

8. "Speech Gene" a Debut Timed to Modern Humans,'" http://www.apnet.com.

9. "'Speech," http: //www.apnet.com.

10. Aitchison, *Speech*; 148. Cite P. Bloom, "Generativity within Language and Other Cognitive Domains," *Cognition*, 51, 1994, 181.

11. Aitchison, 50.

12. Aitchison, 4.

13. Dunham, Will, "Dinosaurs Survived Cataclysm 200 Million Years Ago," *The New York Times*, Thursday, May 10, 2001.

14. Aitchison, 62.

15. Aitchison, 119. See also George Lakoff, "Cognitive Versus Generative Linguistics: How Commitments Influence Results," *Language and Communication*, 11 (1991), 53–62, 58.

16. Gell-Mann, Murray, "What is Complexity?" *Complexity*, Vol 1, No 1, (John Wiley & Sons, Inc., 1995), 16.

17. Gell-Mann, *Complexity*, 9.

18. Gell-Mann, *Complexity*, 9.

19. Cowan, George, email to author, Thursday, May 31, 2001, from the Santa Fe Institute, New Mexico.

20. Cowan, email.

21. Mishkin, Mortimer, and Appenzeller, Tim, "The Anatomy of Memory," *Scientific American* off-print, 1987.

22. Mishkin, *Memory,* 7.

23. Mishkin, *Memory,* 7.

24. Mishkin, *Memory,* 10.

25. Althaus, Jean-Pierre, *Voyage dans le théâtre,* (Lausanne, Editions Pierre Marcel Favre, 1984), 167.

26. Smith, Anna Deveare, as cited by Kyle Brenton, *A.R.T. News.* Vol XXII, No. 1 (November, 2000) 18.

27. Hovers, Erella, et al, *Neandertals of the Levant,* Institute for Human Origins, Berkeley, January/February 1996, 49, 50.

28. Lorenz, Malkus, Spiegel, Farmer, et al, "Deterministic Nonperiodic Flow," *Journal of Atmospheric Sciences,* 20 (1963) 130–41.

29. Kauffman, Stuart, *At Home in the Universe: The Search for Laws of Self Organization and Complexity,* (New York: Oxford University Press, 1995).

30. Armstrong, Gordon, "Cultural Politics and the Irish theatre: Samuel Beckett and the New Biology," *Theatre Research International,* Vol. 18, No. 3. 215–21: "The fact is that when we examine evolution, it is immediately apparent that the figure of an actor appearing on-stage in a performance arena is only one tiny fragment of an epic story whose likelihood is manifestly unreasonable." 216.

31. Lloyd, Seth, "Learning How to Control Complex Systems," *Bulletin,* The Bulletin of the Santa Fe Institute, Spring 1995, Vol. 10 No 1, 18.

CHAPTER TWO: COMPLEX SOLUTIONS

1. Ross, Phillip, "Hard Words," *Scientific American,* Vol. 264, No. 4 (April, 1991), 140.

2. Ross, "Words," 142–3.

3. Bickerstein, Derek, cited in Ross, "Words," 146.

4. Lieberman, Philip, *Uniquely Human: The Evolution of Speech, Thought, and Selfless Behavior,* (Harvard: Harvard University Press, 1991), 76.

5. It is also noteworthy that Sumerians are credited with inventing the wheel, the arch, and with establishing a system of numerical notations related to the base of sixty, our own basis of time: sixty seconds in a minute, sixty minutes in an hour; and direction: 360 degrees of a compass. See Carter, Martha L., and Keith Schoville, eds., *Sign, Symbol, Script: An Exhibition on the Origin of Writing and the Alphabet,* (Madison: University of Wisconsin Press, 1984),12f.

6. Chomsky, Noam, cited in Ross, "Words," 147.

7. Heidegger, Martin, cited in Stephen Erickson, *Language and Being: An Analytical Phenomenology* (New Haven: Yale U. Press, 1970), 39.

8. Erickson, *Language,* 39–40.

9. Recer, Paul, Associated Press Science Writer, Zhe Xi Luo "Tiny Animal May Be Mammals' Ancestor," 24 May, 2001.

10. Luo, "Ancestor," 2001.

11. The German scholar, Wilhelm Dilthey, proposed an end to the positivist school of theatre, based on principles of natural science at Antoine's Théâtre Libre. In its place, Dilthey advocated that scholars in the field of humanities "should focus on the individual work which can only be understood by experiencing it."

12. Madison, G. B., *The Hermeneutics of Post modernity: Figures and Themes.* (Bloomington: *Indiana U. Press,* 1988), 155 ff.

13. Fichback, "Mind," 56.

14. Edelman, Gerald, *Neural Darwinism: The Theory of Neuronal Group Selection.* (New York: *Basic Books, Inc.,* 1987), 6–7.

15. See David Hellerstein, "Plotting a Theory of the Brain," *The New York Times Magazine,* 23 May, 1988. Interview and review of Edelman *Neural Darwinism.* 61.

16. Wilson, Robert, "Theaterproben IX Rehearsals," *The Forest* Playbook, n.p. (Wilson described the confrontation of the moment of their walking toward each other as "too complex" for explanation,) ["To explain would be to destroy the moment,"]

17. Argyros, Alexander, *A Blessed Rage for Order: Deconstruction, Evolution, and Chaos.* Ann Arbor: *University of Michigan Press,* 1991, i.

18. Jerison, Harry J., *Evolution of the Brain and Intelligence.* New York: *Academic Press,* 1973, 17.

19. Jerison, *Intelligence,* 41, 43.

20. Edelman, *Darwinism,* 3.

21. Rosenfield, Israel, "Neural Darwinism: A New Approach to Memory and Perception," *The New York Review of Books,* Vol xxxiii, No. 15 (October 9, 1986), 21ff.

22. Edelman, *Darwinism,* 210.

23. Goldberg, Elkhonon, *The Executive Brain: Frontal Lobes and the Civilized Mind,* New York: *Oxford University Press,* 2001, 30.

24. Goldberg, *Brain,* 85–86.

25. Gould, Jay, lecture in celebration of inauguration of Columbia University president, Columbia University, October 4, 1993.

26. See discussion in Gordon Armstrong, "Images in the Interstice: The Phenomenal Theater of Robert Wilson," *Modern Drama,* Vol XXXI, 4 (1988): 557–571.

27. Notice should be taken in particular of the split nature of Gilgamesh and Enkidu; their closure would create a figure half god, half animal, which is a poetic way of expressing the nature of man.

28. The divisions, named and indicated by square brackets, are the author's, as are the divisions formulated by sub-headings. For the full text, the reader should consult the Playbook itself.

29. Kuhn, Thomas S., *The Structure of Scientific Revolutions,* 2nd edition (Chicago: *University of Chicago Press,* 1970).

30. Roach, Joseph R., *The Player's Passion: Studies in the Science of Acting* (Newark: *University of Delaware Press,* 1985), 16.

31. Roach, *Passion,* 226.

32. Turner, Victor, "Body, Brain, and Culture; The Anthropology of Performance," *Performing Arts Journal,* 1986, 156.

33. Turner, "Body," 177.

CHAPTER THREE: INTERSTICES

1. Mukarovsky, Jan, "Art as Semiotic Fact," trans. by I. R. Titunik in *Semiotics of Art: Prague School Contributions* (Cambridge: The MIT Press, 1976), 8.
2. Mukarovsky, "Art," 8ff.
3. Oliver, William I., U.C. Berkeley, personal commentary to this writer, 1993.
4. Schechner, Richard, *Performance Theory, Rev. and Exp. Edition*, New York: Routledge, 1988, 6, 146, 270.
5. Hirsch, "Interpretations," commentary on Rene Wellek and Austin Warren, *Theory of Literature*, 2nd edition, New York: Harcourt-Brace, 1956, 31, 465.
6. Hirsch, "Interpretation," 464.
7. Wellek and Warren, *Literature*, 43.
8. MacLean, Paul, "The Triune Brain, Emotion and Scientific Basis," in F. O. Schmidt, ed., *The Neurosciences, Second Study Program*, New York, 1970, 336–349
9. MacLean, Paul, "The Triune Brain," in Robert Isaacson, *The Limbic System* (New York, 1976), 220–221.
10. Isaacson, *Limbic System* , 221.
11. Foucault, Michel, *The Archeology of Knowledge,* trans. A. M. Sheridan Smith (New York, 1972), 29.
12. Foucault, *Archeology*, 29–30.
13. Foucault, *Archeology*, 7, 9.
14. Note that it is the *intentional displacement* that accounts for the recognition factor of a staged action, as opposed to an action in real life. The quality of the action is sustained even though the "figure/ground" relationships are shifted.
15. Picasso, Pablo, *Les Demoiselles d'Avignon*, Paris, June-July 1907. Oil on canvas, 8"-0" x 7'-8". The Museum of Modern Art, New York.
16. Golding, John, "Two Who Made a Revolution: Picasso and Braque: Pioneering Cubism," *Museum of Modern Art*, New York, 1989, 10.
17. Rubin, William, "Picasso and Braque: An Introduction," in Golding, *Picasso and Braque: Pioneering Cubism*, New York: The Museum of Modern Art, 1989, 16.
18. Interesting work was reported on by Susanna Bloch, et al, "Effector Patterns of Basic Emotions," *J. Social Biol. Struct.* 1987 (10) 1–19; and Tadunobu Tsunoda, "Human brain Function and the Culture," Performance and Life Sciences summer course, The Council of Europe, Saintes, France, 24 July-3 August, 1989.
19. Beckett, Samuel, Paris, September 15, 1985.

CHAPTER FOUR: CHAOTICS

1. Glick, James, *Chaos: Making a New Science.* New York: Penguin Books, 1987, 6.
2. Glick, *Chaos*, 261–62.
3. Fichback, Gerald D., "Mind and Brain, " *Scientific American*, September, 1992, 48.
4. McAuliffe, Kathleen, "Get Smart: Controlling Chaos," *Omni*, Vol. 12, No. 5 February, 1990, 44.
5. Fichback, "Mind," 49.

6. Glick, *Chaos*, 235.

7. A Mandelbrot set is a computer generated image of the boundary between two or more attractors in a dynamical system. Each attractor in a system has its own basin. Each basin has its own boundary. The study of fractal basin boundaries is the study of systems that can reach one of several nonchaotic final states.

8. Jones, Edwin, *Reading the Book of Nature. A Phenomenological Study of the Creative Expression in Science and Painting*, Athens: *Ohio U. Press*, 1989. 30.

9. Woolf, Virginia, "Modern Fiction," in *The Common Reader First Series*. London: *Hogarth Press.*, repr. 1975. 189.

10. Weinberg, Stephen, "Can Science Explain Everything? Anything?" *NYRB* Vol. XLVIII, No. 9 (May 31, 2001) 47–50.

11. Hayles, Katherine N., *Chaos and Order: Complex Dynamics in Literature and Science*. Chicago: *The University of Chicago Press*, 1991, 5.

12. NOVA, "The Strange New Science of Chaos," PBS production, WGBH, January 31, 1989. Copyright 1989/*WGBH Educational Foundation*, 4.

13. Chen, Ping, "Nonequilibrium and Nonlinearity: A Bridge Between the Two Cultures," in Scott, George P., Ed., *Time Rhythms ,and Chaos in the New Dialogue with Nature*. Ames: *Iowa State University Press*, 1991, 70.

14. Chen, Chaos, 71.

15. Weinberg, 50.

16. Email to author from Andi Sutherland, editorial staff member of Santa Fe Institute, May 30, 2001.

17. *Santa Fe Institute, SFI Bulletin*, Fall 2000, Vol. 15, No. 2, inside cover page.

18. Chen, "Nonequilibrium," 78.

19. Chen, "Nonequilibrium," 83.

20. Argyros, Alexander J., *A Blessed Rage for Order: Deconstruction, Evolution, and Chaos*. Ann Arbor: The *University of Michigan Press*, 1991. 4.

21. Argyros, *Order*, 170.

22. See Armstrong, "Cultural Politics and the Irish Theatre: Samuel Beckett and the New Biology," *Theatre Research International*, Vol 18, No. 3, 1993. 215–221.

23. Artaud, Antonin, "To Paul Thévenin," Tuesday 24 February 1948, cited in Schumacher, *Artaud,*, 199.

24. Begley, Adam, "The Tempest around Stephen Greenblatt," *The New York Times Magazine*, March 28, 1993, 36.

25. Argyros, *Order*, 7.

26. Artaud, "Chancellors" 17.

27. Nietzsche, Friedrich, *The Geneology of Morals*, in *The Birth of Tragedy and the Geneology of Morals*, trans. Francis Golffing, New York: *Doubleday & Co.*, 1956, 298–99.

28. Artaud, Antonin, "To Have Done with the Judgment of God," a radio play, from "Les Voixes des Poétes," banned the day before broadcast, 2 February, 1948. Readers included Artaud, Roger Blin, Maria Casarès, and Paul Thévenin. See Sontag, Susan, *Antonin Artaud: Selected Writings*, 1976, 555–571, "Notes," 658–60.

29. Artaud. *Judgement of God*: "Conclusions," 570–71.

30. Carlson, Marvin, *Theories of the Theatre: A Historical and Critical Survey, from the Greeks to the Present. Expanded Edition,* Ithaca, N. Y.: *Cornell University Press,* 1993, 262.

31. If we consider Nietzsche's philosophy as based on classical Greek Apollonian and Dionysian principles, and Artaud's deep-structured aesthetic as looking towards the future present, there is solid scientific evidence to suggest that—at least in his understanding of the inner compulsion of man as a theatrical species— Artaud had opened the doorway to performance precepts that may well extend into the 21st century.

32. See the work of the Santa Fe Institute's Complex Systems research in the last decade, and the publications and video tapes that are available for researchers in the field: Santa Fe Institute, 1399 Hyde Park Road, Santa Fe, New Mexico 87501.

33. Schumacher, Claude, ed., "Artaud and Theatre," in *Artaud on Theatre,* London: Methuen, 1989, xxiii.

34. Artaud, Antonin, *To Have Done With the Judgment of God (1947):*

> Kré
>
> kré
>
> pek
>
> kre
>
> e
>
> pte
>
> > ...o recho modo
> >
> > to edire
> >
> > di za
> >
> > tau dari
> >
> > do padera coco

35. Sontag, Susan, *Antonin Artaud: Selected Writings,* Ed, and With an Introduction by Susan Sontag, New York: *Farrar, Straus and Giraux,* 1976, 555, 560.

36. Sontag, *Artaud,* 570–71.

37. Knapp, Bettina, *Antonin Artaud: Man of Vision,* New York: *Avon Books,* 1969, 225.

38. Sontag, *Artaud,* lix.

39. Carlson, Marvin, *Theories of the Theatre: A Historical and Critical Survey, from the Greeks to the Present,* Expanded Edition, 1993. New York: *Cornell U. Press,* 1993. 454.

40. Sontag, *Artaud,* 569.

41. Carlson, *Theories,* 491–500, and elsewhere in this important resummation of Twentieth-century dramatic theories of the theatre.

42. Lorenz, Edward, *Journal of the Atmospheric Sciences, 20* (1963) 130–41. Edward Lorenz and his Attractor model, with its deterministic nonperiodic flow, and his conjecture that all natural systems are governed by "sensitive dependence on initial conditions," began the computer-simulations postmodern era.

43. Warren Weaver, inventor in 1948 of the term "complexity" in an article in the *American Scientist,* is cited by George Cowan as the first person to realize the growing importance of the subject to the scientific community worldwide.

44. Carlson, *Theories,* 497.

45. Johansen, J. Dines, et all, *Approaching Theatre.* Bloomington: *Indiana U. Press,* 1991. 100.

46. Sontag, *Artaud,,* xxxix.

47. Carlson, Marvin, in Helbo, *Approaching Theatre*, 49.

CHAPTER FIVE: CONSCIOUSNESS AND CRAFT

1. Copeland, Aaron, *Music and Imagination*, Cambridge: Harvard U. Press, 1952, 43.
2. Dreyfuss, Richard, *New School* televised interview, January 4, 2001.
3. Eccles, Sir John, *Molecular Neurobiology of the Mammalian Brain*, New York: Routledge, 1978, p. 499.
4. Gershon, Eliot, and Rieder, Ronald D., "Major Disorders of Mind and Brain," *Scientific American*, September, 1992, 129–30.
5. Schumacher, Claude, ed. and trans., and Singleton, Brian, trans, *Artaud on Theatre*, London: Methuen, 1989.
6. Artaud, Antonin, "To the Directors of the Comédie Française," Schumacher, *Artaud on Theatre*, 15–16, First published in *84, No. 13, March 1950*.
7. Artaud, Antonin, see letters to Charles Dullin, 1921; Max Jacob, October, 1921; Mademoiselles Yvonne Gillies, October, 1921; letters on L'Atelier Theatre, 1922; *Six Characters in Search of an Author*, 1923; Maurice Maeterlinck, 1923; and publication of *The Evolution of Set Design*, 1924; with full citations in Schumacher, *Artaud*, 3ff.
8. McAuliffe, "Smart," 87.
9. McAuliffe, "Smart," 88ff.
10. Knapp, Bettina, *Antonin Artaud: Man of Vision*, New York: Avon Books, 1969, 60.
11. Knapp, *Artaud*, 60.
12. Schumacher, *Artaud*, xxx, Citation from Marowitz, Charles, *Artaud at Rodez*, London: Marion Boyars, 1977, 108,
13. Sontag. *Artaud*, lvii.
14. Artaud, in Schumacher, *Artaud*, 16.
15. Sophocles, *Oedipus Rex*, trans. Richard Lattimore, Chicago: U. of Chicago Press, offers the most dramatically lucid play-text.
16. Kinney, Hannah, et all, "Neuropathological Findings in the Brian of Karen Ann Quinlan: The Role of the Thalamus in the Persistent Vegetative State," *The New England Journal of Medicine*, Vol 20, No 21 (May 26, 1994), 1469–75.
17. Goldberg, Elkhonon, *The Executive Brain: Frontal Lobes and the Civilized Mind*, New York: Oxford U. Press, 2001. See in particular A. R. Luria, (Goldberg's mentor), *Higher Cortical Functions in Man*, New York: Basic Books, 1966.
18. Goldberg, 216.

CHAPTER SIX: COMPLEXITY AND DISCONTINUITY OF SPECIES

1. Burkhardt, Frederick, ed., *Charles Darwin's Letters; A Selection*, Cambridge U. Press, 1996, 81.
2. Aitchison, *Speech*, 102.
3. Darwin. Charles, *Origin of Species*, New York: P. F. Collier & Son, Corp., 1909, *Species*, 617.
4. Darwin, *Species*, 170.

5. Smith, John Maynard, "The Cheshire Cat's DNA," review of Evelyn Fox Keller, *The Century of the Gene, Harvard U. Press*, in *The New York Review of Books*, Vol. XLVII, No. 20 (December 21, 2000), 43.

6. Aitchison, *Speech*, 134.

7. Nichols, Johanna, "Linguistic Diversity and the First Settlement of the New World," *Language*, vol. 66, 475–521.

8. Aitchison, *Speech*, 165–66.

9. Aitchison, *Speech*, 171.

10. Nichols, *Linguistic Diversity in Space and Time, University of Chicago Press*.

11. Recer, "Ancestor," 2001.

12. Angelou, Maya, President William Jefferson Clinton's Inauguration Day Poem, "On the Pulse of the Morning," January 20, 1993. Reprinted in *The Providence Journal*, January 21, 1993,1.

13. Beckett, Samuel, *The Lost Ones*, (New York: *Grove Press*, 1972), 10–11, 25.

14. See Robert L. Isaacson, *The Limbic System*, New York: *The Plenum Press*, 1974, 236ff.

15. Edelman, Gerald, *Bright Air, Brilliant Fire, On the Matter of the Mind*, New York: *Basic Books*, 1992, 108.

16. Jastrow, *Loom*, 128.

17. Edelman, "Topobiology: Lessons From the Embryo," *Fire*, 53–55. The description that follows is explained in greater detail in Edelman's narrative and diagram, fig. 6–1, 54.

18. Secondary association areas are the key to understanding how human language works. Norman Geschwind proposed that the development of *association areas* of the *inferior parietal lobule* freed man from the dominant pattern of sensory-limbic associations, and permitted the development of speech, based on a *sound-image* and *concept* dichotomy. See Geschwind, Norman, and Gala-burda, Albert, Eds., "Disconnection Syndromes in Animals and Man," in *Cerebral Dominance: the Biological Foundations*, Cambridge: *Harvard U. Press*, 1984, 107–8.

19. Artaud, Antonin, "To André Gide," Paris, Sunday, 7 August 1932, in Schumacher. Claude, Ed., *Artaud on Theatre*, London: *Methuen*, 1991. 68–9.

20. Fischer-Lichte, Erica, "Theatre Historiography and Performance Analysis: Different Fields—Common Approaches?" *Assaphi*, Section C, No. 10, 1994, 99.

21. Carlson, Marvin, "Indexical Space in the Theatre," *Assaphi*, 9

22. Lyons, Charles, "Character and Theatrical Space," in *Theatrical Space: Themes in Drama 9*, ed. Redmon, James. London: *Cambridge U. Press*, 1987, 36.

23. Carlson, "Indexical," 9.

CHAPTER SEVEN: NATURE'S RETURN

1. Kauffman, Stuart, *At Home in the Universe: The Search for Laws of Self Organization and Complexity*, New York: *Oxford University Press*, 1995.

2. Armstrong, Gordon, Cultural Politics and the Irish theatre: "Samuel Beckett and the New Biology," *Theatre Research International*, Vol. 18, No. 3. 215–21. "The fact is that when we examine evolution, it is immediately apparent that the figure of an actor appearing onstage in a performance arena is only one tiny fragment of an epic story whose likelihood is manifestly unreasonable." 216.

3. I believe the answer is "Yes." Given the need to adapt or perish across millenniums, the genus from which we are descendent survived the climatic and geologic crises, where most others disappeared.

4. Brockett, Oscar, *History of Theatre*, 6th Edition. Boston: *Allyn and Bacon*, 1991. 1.

5. Sontag, Susan, in Leslie Garis, "Susan Sontag Finds Romance," *The New York Times Magazine*, August 2, 1992, Section 6, 43.

6. Lorenz, *Atmospheric Sciences*, 130–41.

7. I will not elaborate further, except to remind readers to review the sections on neural development of man from the perspective of Princeton University psychologist Robert Jaspers. This point has been dealt with in the text.

8. Limits to this approach have recently been expressed by George Cowan, founder of the Santa Fe Institute, in distinguishing between "ordered" and "disordered" complexity—the former being a prime target of complexity modeling, and a major concern of science. Correspondence, May 31, 2001.

9. Cowan, correspondence.

10. The argument is not so much that these brain areas developed at this time but that latent possibilities emerged as connections, made to the *prefrontal lobes* of the *cerebral cortex* from the *angular gyrus* and the *mid brain*, fostered their integration into patterns that we now call speech and consciousness. The possibilities may have lain dormant for millions of years.

11. Lloyd, Seth, "Learning How to Control Complex Systems," *The Bulletin of the Santa Fe Institute*: Spring 1995, 20.

12. Lloyd, *Bulletin*, 19.

13. Barthes, Roland, *Tel Quel*, 1963, .

14. Carlson, Marvin, "Indexical Space in the Theatre," in "Theatrical Space and Fictional Space," *Assaphi: Studies in the Theatre*, Section C, No. 10. ed. Rozik, Eli, 1.Kott, Jan, "The Icon and the Absurd," *The Drama Review*, 14, 1969, 19.

15. Carlson, "Indexical," 1

16. Blau, Herbert, "Universals of Performance, or Amortizing Play," *Sub-Stance*, 37–38, 1983, 1557.

17. Carlson, "Indexical," 4.

18. Carlson, "Indexical," 8. See Lyons, Charles, "Character and Theatrical Space," *Theatrical Space: Themes in Drama 9*, ed. Redmon, James, London: *Cambridge U, Press*, 1987, 36.

19. Fischer-Lichte, Erika, "Theatre Historiography and Performance Analysis," *Assaphi*. 10. 106,111.

20. Carlson, "Indexical," 9.

21. Gell-Mann, Murray, "Complex Adaptive Systems," in *The Mind, the Brain, and Complex Adaptive Systems*. ed. Morowitz, Harold, Santa Fe Institute Studies in the Sciences of Complexity; Reading, MA: *Addison-Wesley Publishing Co.*, 1995, 12.

CHAPTER EIGHT: THE DISCONTINUOUS CORTEX

1. Darwin, Charles, quoted in *The Beagle Record*, ed. Richard Darwin Keynes, Cambridge: *Cambridge U. Press*, 1979, 32.

2. Darwin, *Beagle*, 34.

3. Darwin, Charles, *Autobiography*, ed. Gavin de Beer, London: *Oxford University Press*, 1974, 71.

4. Darwin, *Species*, 137.

5. Darwin, *Autobiography*, 52.

6. Beckett, Samuel, *How It Is*,New York: *Grove Press, Inc.*, 1964, 134.

7. Nietzsche, Friedrich, *Thus Spoke Zarathustra*, trans. Walter Kauffmann, New York: *The Viking Press*, 1966, 13.

8. Nietzsche, Friedrich, *Selected Letters of Friedrich Nietzsche*, ed., and trans., Christopher Middleton, Chicago: *University of Chicago Press*, 1969, 208.

9. Nietzsche, *Zarathustra*, 17.

10. Nietzsche, *Selected Letters*, 209.

11. Artaud, Antonin, *The Theatre and Its Double*, trans. Mary Richards, New York, 1958, 30.

12. Artaud, *Double*, 26–27.

13. Nietzsche, *Zarathustra*, 216.

14. Bogatryev, Pietr, "Semiotics in the Folk Theatre," 1938, transl. and reprinted in Ladislaw Matejka and Irwin Titunuk, *Semiotics of Art: Prague School Contributions*, Cambridge: *Cambridge U. Press*, 1976), 35–36.

15. Carlson, *Theories*, 514.

16. Blau, Herbert, "Look What Thy Memory Cannot Contain," 1981, in *Blooded Thought*, New York, 1982, 93.

17. Blau, *Thought*, 157.

18. Beckett, Samuel, *The Unnamable*, in *Three Novels by Samuel Beckett*, New York: Grove Press, 1965, 414.

19. Beckett, *The Unnamable*, 414.

20. Beckett, *The Unnamable*, 299.

21. Mamet, David, *Speed-the-Plow*, Goodman Theatre, 1988.

22. Mamet, *The New York Times*, January 20, 1988, C13, 15.

23. Mamet, *The New York Times*, May 4, 1988. C16.

24. Oliver, William I., letter to the author, October, 1988.

25. Simon, Neil and Rabe, David, "The Craft of the Playwright," ed. Samuel G. Freedman and Michael Williams, *The New York Times Magazine*, May 26, 1985, 38, 52.

26. Wlson, Robert, unpublished application for funding of *the CIVIL warS* for the 1984 Los Angeles Olympic Arts Festival, Cambridge, MA., American Conservatory Theatre Archives.

27. Wilson, *warS* funding.

28. Wilson, Robert, Byrns, David, "Workshop V Workshop," *The Forest* Program, *Theater der Freien Volkesbuhne*, (Berlin, 1988), n.p.

29. Harriott, Esther, "Sam Shepard: Inventing Identities," *American Voices: Five Contemporary Playwrights in Essays and Interviews* (Jefferson, N. C.,: McFarland and Co., Inc.,1988). 16. See also review by Gordon Armstrong, *Theatre Research International*, Vol. 14, No. 3 (Autumn, 1989): 309–311.

30. Wright, Elizabeth, "Modern Psychoanalytic criticism," *Literary Theory*, 113.

31. Commentary in this section has been limited to some discussion of the literary theories of the positivists, postpositivists and the New Critics.

32. Harriott, *Voices*, 152.

33. Pirandello, Luigi, *Preface to Six Characters in Search of an Author*, in *Naked Masks: Five Plays by Luigi Pirandello*, trans. Eric Bentley (Dutton, 1952), repr. Weiss, *Drama in the Modern World*, Heath, 1960, 244.

34. Brooks, *Primacy*, 167.

35. See discussion in Gordon Armstrong, *Samuel Beckett, Jack Yeats, W. B. Yeats: Images and Words* (Associated U. Presses, 1989).

36. Christopher Baker, "Behind the Scenes with Abbie Katz," *American Repertory Theatre News*, Vol IX, No. 2 (February 1989) 13.

37. Beckett, Samuel, conversation with the author, Paris, September 16, 1985.

38. Ortolani, Benito, *The Japanese Theatre: From Shamanistic Ritual to Contemporary Pluralism*, Princeton: Princeton University Press, 1995, 277.

39. I am grateful to Kinneret Noy at the University of Pittsburgh for her assistance and generosity in obtaining this information and in sharing her own research with me.

40. Ortolani, 277.

41. Nario, Goda, "On Ankoku Butô", in Susan Blakely, *Ankoku Butô*, Cornell University Press, 1988, 79.

42. Nario, 81.

43. Video tape, courtesy of Kinneret Noy.

44. Osami, Egichi, "My View of Hopp Butô-ha", in Susan Blakely, *Ankoku Butô*, Cornell University Press, 1988, 59.

45. Gould, Jay, lecture in celebration of inauguration of Columbia University President, Columbia University, October 4, 1993.

46. Edelman, *Darwinism*, 6–7.

47. Damasio, Antonio, & Damasio, Hanna, "Brain and Language," Mind and Brain, *Scientific American*, September, 1992, 57.

48. Gore, Albert, *Earth in the Balance*, 1992. 62ff.

49. Angier, Natalie, "Sonatas for Humans, Birds and Humpback Whales," http://www.nytimes.com/2001/01/09science/09MUSIhtml

50. Kelso, J. A. Scott, *Dynamic Patterns: The Self Organization of Brain and Behavior*, MIT Press, 1995.

51. Kelso, *Patterns*, xvii ff.

52. Kelso, *Patterns*, 263.

53. Beckett, Samuel, conversation with author, July 1985.

54. Poincaré, Henri, in Kelso, in *Science and Hypothesis*, 1952, xxiv.

55. Kelso, *Patterns*, 142.

56. Sherrington, C.S., *Man and his Nature*, 1940, in Kelso, *Patterns*, 258ff.

57. One might be hard-pressed to verify the accuracy of the birdsong that emanated from the larynx of Antonin Artaud, but Susan Sontag has recorded this version of portions of the speech, "To have Done with the Judgment of God," delivered in 1947. Artaud, *Selected Writings*, 1976, 555.

CHAPTER NINE: ANCIENT AND MODERN THEATRE CONSCIOUSNESS

1. Armstrong, Gordon, "Theatre as a Complex Adaptive System," *New Theatre Quarterly*, Vol. XIII, No. 51, August 1997, 277–289.

2. Joseph Roach, "Problems and Prospects for Theatre Research," Statement to the American Council for Learned Societies, undated, in response to a charge from Milton Gatch, to ACLS Delegates on September 11, 1992.

3. Rouse, John, in Janelle Reinelt, "Semiotics and Deconstruction: an Introduction," *Critical Theory and Performance*, eds. Janelle G. Reinelt and Joseph Roach, Ann Arbor: *The University of Michigan Press*, 1992, 113.

4. Report cited in *The Newport Daily News*, Tuesday, January 5, 1993, A3.

5. See Rose Lee Goldberg, *Performance Art: From Futurism to the Present*, New York: *Harry N. Abrams Inc.*, 1988, 153.

6. Stone, Lawrence, "The Revolution Over the Revolution," *New York Review of Books*, Vol.XXXIX, No. 11, June 11, 1992, 47.

7. Smith, John Maynard, "Taking a Chance on Evolution," *New York Review of Books*, Vol XXXIX, No. 9, May 14, 1992, 34.

8. Lewontin, R.C. "The Dream of the Human Genome," *The New York Review of Books*, Vol XXXIX, No. 10 (May 28, 1992), 31.

9. Schrodinger, Erwin, *What Is Life? The Physical Aspect of the Living Cell*, Cambridge: *Cambridge University Press*, 1944, repr. 1955.

10. Knowlson, James, catalogue: *Samuel Beckett: An Exhibition*: Reading University Library, May-July 1971. London: *Turret Books*, 1971, 52.

11. Damasio, Antonio R. and Damasio, Hanna, "Brain and Language," Mind and Brain, *Scientific American*, September, 1992, 89.

12. Smith, Maynard, "Taking a Chance on Evolution," *New York Review of Books*, Vol. XXXIX, No 9, May 14, 1992, 34.

13. Wheeler, John A., cited in Horgan, John, "Quantum Philosophy," *Scientific American*, July 1992. 101.

14. Beckett, Samuel, *Imagination Dead Imagine*, London: *Calder and Boyars*, 1965, 10

15. Beckett, *Imagination*, 11.

16. Schrodinger, *Life*, 51.

17. Schrodinger, *Life*, 69–70.

18. Schrodinger, *Life*, 70.

19. Schrodinger, *Life*, 72.

20. In *The New York Times* of Sunday, February 7, 1993, Section 1, 1, 14, a list appeared of forty-eight current ethnic wars, with a summary of conditions that led to the insurrections, in Europe (9), Middle East and North Africa (7), Africa South of the Sahara (15), Asia (13), and Latin America (4). At that time, who could predict the 21st century escalations by extremists on the fringes of potential world-wide religious confrontations.

21. Beckett, Samuel, *Endgame*, New York: *Grove Press*, 1958, 1.

22. Schrodinger, *Life*, 91.

23. See Bruce Chatwin, *The Songlines*, New York: Penguin Books, 1987; Harvey Arden, *Dreamkeepers: A Spirit-Journey into Aboriginal Australia*, New York: Harper-Collins, 1994.

24. Geertz, Clifford, "Blurred Genres: The Refiguration of Social thought," in *The American Scholar* 49 (1980), reprinted in *Critical Theory since 1965*, ed. Hazard Adams and Leroy Searle (Tallahassee: *Florida State University Press*, 1986), 514–23. cited in *Critical Theory and Performance*, eds. Janelle G. Reinelt and Joseph Roach, Ann Arbor: *The University of Michigan Press*, 1992, 2.

25. See G. B. Madison, *The Hermeneutics of Postmodernity Figures and Themes* Bloomington: *Indiana U. Press*, 1988, xi ff.

26. Lewontin, "Genome," 32.

27. Lewontin, "Genome," 33.

28. Lewontin, "Genome," 32–33.

29. Lewontin, "Genome," 32.

30. Beckett, Samuel, *How It Is*, New York; *Grove Press*, Inc. 1964, 15–16.

31. Beckett, *How It Is*, 30.

32. Beckett, *How It Is*, 30.

33. Beckett, *How It Is*, 38.

34. Beckett, *How it Is*, 47.

35. Beckett, *How It Is*, 48.

36. Beckett's dramaticule, *Not I*, premiering at Lincoln Center in 1971, once again represented Beckett's work at its best. Beginning with an apocryphal story of an Arab Women in Algeria shrieking at passers-bye at a bus stop, which Beckett personally witnessed, the broken social codes became the basis of a deeper exploration of bio-codes that Beckett had explored, in *Comment C'est*, [*How It Is*] six years earlier.

37. Schrodinger, *Life*, 91–2.

38. Beckett, *How It Is*, 109.

39. Beckett, *How It Is*, 112.

40. Schrodinger, *Life*, 31.

41. Beckett, Samuel, *Endgame*, New York: *Grove Press, Inc.*, 1958. Subsequent references in this paper will refer to pages in the printed text.

42. Beckett, Samuel, *Proust*, New York: *Grove Press, Inc.*, 1957.

43. On a visit to Paris in 1980, two days before the election of Ronald Regan to the White House, I spoke to Mr. Beckett about the death of a colleague weeks earlier, and asked him if he could recommend a passage from his work. His response was to quote this passage verbatim to me. [On the prospects for re-election of Carter, Beckett's response was revealing.["If he didn't smile so much...."]

44. Schrodinger, *Life* , 92.

45. Armstrong, Gordon, "Unintentional Fallacies," *Journal of Dramatic Theory and Criticism*, Vol. VII, No. 2, 1992, 7–26.

46. "The Infant Universe, in Detail," Scientific American.com. on February 11, 2003, NASA reported the Wilkinson Microwave Anisotropy Probe (WMAP), a million miles from earth, revealed that the lights in our universe came on only 200 million years after the big bang, 13.7 billion years ago. Furthermore, the universe is flat and will go on forever, as time is measured on earth.

47. Gould, Jay, Columbia University Lecture, October 7, 1993

48. Calvin, William H., *The Ascent of Mind:: Ice Age Climates and the Evolution of Intelligence*, (New York: Bantam Books, 1991), xvi.

49. Lest we pass this discussion of climate off as an aberration out of the past that can have no bearing on our own future, consider some recent evidence produced by Dr. Henry Weiss at Yale University. Climatic changes ended the great Mayan civilization in Mexico and parts of Central America in the ninth century; dramatic drops in temperature and a sudden drought led to the demise of the hunting and gathering Natufian communities of southwest Asia-Middle East in the millennium between 12,500 and 11.500 years ago. Consider a nuclear winter that is a natural happening, lasting for several decades, and we can imagine the consequences on a global scale.

50. Chatwin, *Songlines*, 72.

51. Beckett, Samuel, *How It Is*, (New York: Grove Press, 1965), 134.

52. Kelso, *Patterns.*, 15.

53. Shakhar, Indu, *Sanskrit Drama: Its Origin and Decline*, (Leiden: E.J. Brill, 1960), xiii.

54. Shakhar, *Sanscrit*, xvii.

55. Keith. A. B., *The Sanskrit Drama in its Origin, Development Theory and Practise*, (Oxford: Oxford U. Press, 1924), 16.

56. Keith, *Origin*, 45.

57. Varadpanda, M. L., *Traditions of Indian Theater*, (New Delhi: Abhinar Publications, 1978), 54.

58. Varadpanda, *Traditions*, p. 57. Varadpanda suggests a treatise on dramaturgy, written in the second century B.C. suggesting that playhouses be erected "like a mountain cavern" (54), presumably similar to excavated cave theaters like the Sitabengara cave in the Rangesh hills (circa 300 B.C).

59. Varadpanda, *Traditions*, 77.

60. Shakhar, *Sanskrit*, 58.

61. Shakhar, *Sanskrit*, 61.

62. Hsiang-Kuang, Chou, *The History of Chinese Culture*, (Allahabad: Central Book Deposit, 1960), 3.

63. Garnet, Jacques, *Ancient China: From the Beginning to the Empire*, trans. from the French by Raymond Rudoff, (Berkeley: University of California Press, 1968), 48.

64. Garnet, *China*, 30.

65. Garnet, *China*, 33.

66. Chang, K. C., *Art, Myth and Ritual; The Path to Political Authority in Ancient China*, (Cambridge: Harvard University Press, 1983), 112–114.

67. Victor Mair, quoted by John Noble Wilford, "In Ruin, Symbols on a Stone Hint at a Lost Asian Culture," *The New York Times*, 13 May, 2001.

68. Hsiang-Kuang, *Culture*, 3.

69. Chang, K. C., *Art*, 2.

70. Scott, A. C., *The Classical Theater of China*, (Westport: Greenwood Press, 1957), 28.

71. Garnet, *China*, 48.

72. Old, Walter Gorn, *The Shu King, or the Chinese Historical Classic*, (London: Theosophical Publishing Society, 1904), vi.

73. Chang, *Art*, 1.

74. Chang, *Art*, 51.

75. Chang, *Art*, 52–3.

76. Chang, *Art*, 53.

77. Dolby, William, *A History of Chinese Drama*, (London: Paul Elak, 1976), 2–3.

78. Dolby, *Drama*, 3.

79. Dolby, *Drama*, .2.

80. *Origins of Chinese Drama*, 7; mention of this incident is also made in Dolby, *Drama*, 8.

81. Dolby, *Drama*, 14.

82. Performed at Edfu, circa 1284 B.C., this ceremonial drama was a celebration of Isis, Osiris and his posthumous son, Horus, and the punishment of Seth, who had murdered his brother Osiris, thereby dividing the country into Upper and Lower Egypt. The Pharaoh was the god who re-united the two crowns of Upper [Seth] and Lower [Horus] Egypt by restoring the patrimony of Horus. In the Ptolemaic Period the vanquishing of Seth became a symbol of Egypt triumphing over its occupiers, the best known being Cleopatra, the last in a line of Ptolemys. It is interesting that these legends date from only 2400 B.C, and not earlier.

CHAPTER TEN: TWENTY-SIX STRINGS OF CONSCIOUSNESS

1. Beckett, Samuel, *How It Is*, (New York, Grove Press, 1964). Subsequent page references are to this edition.

2. See exhibit "Skulls," at the California Academy of Sciences, San Francisco Golden Gate Park, in Wattis Hall, for a graphic display of sea lion skulls, transformed by environmental stresses.

3. Weiss, Henry, report by Anne McIlroy, *The Toronto Globe and Mail*, Tuesday. January 30, 2001, A2.

4. Weiss, Rick, "Life's Blueprint in Less Than an Inch," http:// www.washingtonpost.com/wp-articles/A54653–2001Feb10 html, 3 of 6.

5. Waterston, Robert, cited in Weiss, *"Blueprint,"* 5 of 6.

6. Rees, Martin, *Just Six Numbers*: The Deep Forces that Shape the Universe, New York: Basic Books, 2000. 45–6.

7. Rees, Numbers, 46.

8. Rees, 83.

9. Weiss, "Blueprint."

10. For further information, see website: http—zebu.uoregon-js-21ˢᵗ__centu...; under "string theory" heading.

11. Rees, 130.

12. Rees, 84, 101. For more readings on current cosmological theory, see the following: Rocky Kolb, *Blind Watchers of the Sky* (Oxford U. Press); Steven Weinberg, *The First Three Minutes* (Basic Books); David Park, *The Fire Within the Eye* (Princeton U. Press); Mario Livio, *The Accelerating Universe* (John Wiley). I am also grateful to Professor Frank Levin, particle physicist at Brown University, for his lecture series on this topic at Salve Regina University, Fall, 2002.

Bibliography

Althaus, Jean-Pierre, *Voyage dans le théâtre*, Lausanne, Editions Pierre Marcel Favre.

Antoine, André, "Letter to Francisque Sarcey," Adolphe Thalasso, Le Théâtre Libre, trans. unkn., Paris, 1909.

Arendt, Hannah, *The Life of the Mind*, San Diego: Harcourt Brace Jovanovich, 1978.

Argyros, Alexander J., *A Blessed Rage for Order: Deconstruction, Evolution, and Chaos*. Ann Arbor: The University of Michigan Press, 1991.

Armstrong, Gordon, "Cultural Politics and the Irish theatre: Samuel Beckett and the New Biology," *Theatre Research International*, Vol. 18, No. 3.

———, "Images in the Interstice: The Phenomenal Theater of Robert Wilson," *Modern Drama*, Vol XXXI, 4, 1988.

———, "Samuel Beckett and the New Biology," *Theatre Research International*, Vol 18, No 3, Fall, 1993.

———, *Samuel Beckett, Jack Yeats, W. B. Yeats: Images and Words*, Associated U. Presses, 1989.

———, "Theatre as a Complex Adaptive System," *New Theatre Quarterly*, Vol. XIII, No. 51, August 1997.

———, "Unintentional Fallacies," *Journal of Dramatic Theory and Criticism*, Vol. VII, No. 2, 1992.

Arnott, James, "An Introduction to Theatrical Scholarship," *New Theatre Quarterly*, 39 (1981).

Artaud, Antonin, "To André Gide," Paris, Sunday, 7 August 1932, in Schumacher. Claude, Ed., *Artaud on Theatre*, London: Methuen, 1991.

———, "To Paul Thévenin," Tuesday 24 February 1948, cited in Schumacher, ed. and trans., and Singleton, Brian, trans, *Artaud on Theatre*, London: Methuen, 1989.

———, "To the Directors of the Comédie Française," Schumacher, Claude, ed., trans. Schumacher, Claude, and Singleton, Brian, *Artaud on Theatre*, London: Methuen House, 1989, 15–16, First published in *84*, No. 13, March 1950.

Auden, W. H., *The English Auden: Poems, Essays and Dramatic Writings 1927–1939*, Ed., Edward Mendelson, London: Faber and Faber, 1977.

Baker, Christopher, "Behind the Scenes with Abbie Katz," *American Repertory Theatre News*, Vol IX, No. 2, February 1989.

Barthes, Roland, *Elements of Semiology*, trans. A. Lavers and C. Smith, London, 1967.

Basnell, Susan, review of John Hilton, *Performance*, in *Theatre Research International* 14 (1989).

Beckett, Samuel, *Endgame*, New York: Grove Press, 1958.

———, *How It Is*, New York: Grove Press, Inc. 1964.

———, *Imagination Dead Imagine*, London: Calder and Boyars, 1965.

————, *The Lost Ones*, New York: Grove Press, 1972.

————, *The Unnamable*, in *Three Novels by Samuel Beckett*, New York: Grove Press, 1965.

————, *Watt*, New York: Grove Press, Inc., 1959.

————, *How It Is*, New York: Grove Press, Inc., 1964.

Begley, Adam, "The Tempest around Stephen Greenblatt," *The New York Times Magazine*, March 28, 1993.

Benedikt, Michael, *Modern French Theatre*, New York: E.P. Dutton, 1964.

Bernard, Claude, *An Introduction to the Study of Experimental Medicine*, trans. Henry Green, New York: Dover Publications, 1957.

Blau, Herbert, "Look What Thy Memory Cannot Contain," (1981), in *Blooded Thought* (New York, 1982).

Blau, Herbert, "Universals of Performance, or Amortizing Play," *Sub-Stance, 37–38*, 1983.

Bloch, Susanna, et al, "Effector Patterns of Basic Emotions," *J. Social Biol. Struct.* 1987 (10) 1–19; and Tadunobu Tsunoda, "Human Brain Function and the Culture," Performance and Life Sciences summer seminar, The Council of Europe, Saintes, France, 24 July-3 August, 1989.

Bogatryev, Petr, "Semiotics in the Folk Theatre," (1938), transl. and reprinted in Ladislaw Matejka and Irwin Titunuk, *Semiotics of Art: Prague School Contributions*, Cambridge, 1976.

Brissett, Dennis, and Charles Edgley, Life as Theater: A Dramaturgical Sourcebook, 2nd Edition, New York: Aldine de Gruyter, 1990.

Brooks, Cleanth, "The Primacy of the Linguistic Medium," *The Missouri Review*, Vol. 6, No. 3, Summer, 1983.

Brooks, Cleanth, "The Primacy of the Reader," *Missouri Review*, Vol 6, No. 2, Winter 1983..

Büchner, Georg, *Woyzeck*, in Corrigan, Robert, *The Modern Theatre*. New York: The Macmillan Company, 1964.

Calvin, William H., *The Ascent of Mind: Ice Age Climates and the Evolution of Intelligence*, New York: Bantam Books, 1991.

Canfield, William, *Francis Picabia*, Princeton, NJ: Princeton University Press, 1987.

Canguillo, Francisco, *Le Serate Futurists*, Naples, 1930, Rpt., Milan Cara Editrice Ceschina, 1961, in *Marinetti: Selected Writings*, 1974.

Carlson, Marvin, "Indexical Space in the Theatre," in "Theatrical Space and Fictional Space," *Assaphi, Studies in the Theatre*, Section C, No. 10. Ed. Rozik, Eli.

Carlson, Marvin, *Theories of the Theatre: A Historical and Critical Survey, from the Greeks to the Present*, Expanded Edition, 1993. New York: Cornell U. Press, 1993.

Carlson, Marvin, "Towards a New Historiography," ASTR Meeting, New York, November, 1985.

Carr, David, *Phenomenology and the Problems of History*, Evanstown, Ill.: Northwestern U. Press, 1974.

Carter, Martha L., and Keith Schoville, eds., *Sign, Symbol, Script: An Exhibition on the Origin of Writing and the Alphabet*, Madison: University of Wisconsin Press, 1984.

Cézanne, Paul, cited in Edwin Jones, *Reading the Book of Nature*, Athens: Ohio U, Press, 1985.

Chang, K. C., *Art, Myth and Ritual: The Path to Political Authority in Ancient China*, Cambridge: Harvard University Press, 1983.

Chen, Ping, "Nonequilibrium and Nonlinearity: A Bridge Between the Two Cultures," in Scott, George P., Ed., *Time Rhythms, and Chaos in the New Dialogue with Nature*, Ames: Iowa State University Press, 1991.

Copeland, Aaron, *Music and Imagination*, Cambridge: Harvard U. Press, 1952.

Culler, Jonathan, *On Deconstruction: Theory and Criticism after Structuralism*, Ithaca: Cornell U. Press, 1982.

Damasio, Antonio R., and Hanna, "Brain and Language," *Mind and Brain, Scientific American*, September 1992.

Darwin, Charles, *Autobiography*, ed. Gavin de Beer, London: Oxford University Press, 1974.

Darwin, Charles, *Origin of Species*, New York: P. F. Collier & Son, Corp., 1909.

de Saussure, Ferdinand, "Course in General Linguistics," London, 1978, trans. from *Cours de linguistique général*, Paris, 1916, in David Robey, "Modern Linguistics and the Language of Literature," *Modern Literary Theory: A Comparative Introduction*, ed. Ann Jefferson and David Robey, Totawa, N.J.: Barnes and Noble, 1984.

Derrida, Jacques, *L'Ecriture at la différence* (Paris, 1967), 152. Eng. trans. *Writing and Difference*, Chicago: U. of Chicago Press, 1978.

Dolby, William, *A History of Chinese Drama*, London: Paul Elak, 1976.

Eccles, John, *Evolution of the Brain: Creation of the Self*, New York: Routledge, 1989.

Eco, Umberto, *A Theory of Semiology*, Bloomington: U. of Indiana Press, 1976.

Edelman, Gerald, *Bright Air, Brilliant Fire: On the Matter of the Mind*, New York: Basic Books, 1992.

Edelman, Gerald, *Neural Darwini, The Theory of Neuronal Group Selection*, New York: Oxford U. Press, 1987.Elam, Keir, *The Semiotics of Theatre and Drama*, London, 1980.

Eldredge, N. and Gould, S. J., "Punctuated Equilibria: An Alternative to Phyletic Gradualism," *Models of Paleobiology*, ed. T.J.M. Schopf, San Francisco: Freeman, Cooper, 1972.

Fichback, Gerald D., "Mind and Brain," *Scientific American*, September 1992.

Fischer-Lichte, Erika, "Theatre Historiography and Performance Analysis: Different Fields— Common Approaches?" *Assaph*, Section C, No. 10, 1994.Fischer-Lichte, Erika, *The Semiotics of Theater*, Trans. Jeremy Gaines and Doris Jones, Bloomington: Indiana U. Press, 1992.

Foucault, Michel, *The Archeology of Knowledge*, Trans. A. M. Sheridan Smith, New York: Harper and Row, 1976.

Freud, Sigmund, *Leonardo da Vinci: A Study in Psychosexuality*, Trans. A. A. Brill New York: Random House, 1967.

Garnet, Jacques, *Ancient China: From the Beginning to the Empire*, trans. from the French by Raymond Rudolf, Berkeley: University of California Press, 1968.

Gazzaniga, Michael, *The Social Brain:: Discovering the Networks of the Mind*, New York: Basic Books, 1985.

Geertz, Clifford, "Blurred Genres: The Refiguration of Social thought," in *The American Scholar* 49 (1980), reprinted in *Critical Theory since 1965*, ed. Hazard Adams and Leroy Searle, Tallahassee: Florida State University Press, 1986, 514–23. Cited in *Critical Theory and Performance*, eds. Janelle G. Reinelt and Joseph Roach, Ann Arbor: The University of Michigan Press, 1992.

Gell-Man, Murray, "Complex Adaptive Systems," in *The Mind, the Brain, and Complex Adaptive Systems*, ed. Morowitz, Harold. Santa Fe Institute Studies in the Sciences of Complexity, Reading, MA: Addison-Wesley Publishing Co., 1995.

Gershon, Eliot, and Rieder, Ronald D., "Major Disorders of Mind and Brain," *Scientific American*, September, 1992.

Geschwind, Norman, and Galaburda, Albert, Eds., "Disconnection Syndromes in Animals and Man," in *Cerebral Dominance: the Biological Foundations*, Cambridge: Harvard U. Press, 1984.

Geschwind, Norman, *Language and the Brain*, Boston.: D. Reidel, 1974.

———, *Selected Papers on Language and the Brain*, Boston: D. Reidel, 1974.

———, *Neural Darwinism: The Theory of Neuronal Group Selection*, New York: Basic Books, 1987.

———, *Bright Air, Brilliant Fire: On the Matter of the Mind*, New York: Basic Books, 1992.

Gide, André, *Oeuvres Complètes*, Paris: *Nouvelle revue francais, 1932–39*.

Gilot, Francois, and Carleton Lake, *Life with Picasso*, New York, 1964.

Glaizer, Albert, "L'Affair Dada," in *Action*, April 20, 1920, 26–3, trans., Ralph Manheim, *The Dada Painters and Poets*, ed., Robert Motherwell, New York: Wittenborn, Schultz, 1951.

Gleick, James, *Chaos: The Making of a New Science*, New York: Penguin Books, 1987.

Goldberg, Rose Lee, *Performance Art: From Futurism to the Present*, New York: Harry N. Abrams Inc., 1988.

Golden, Leon, *Aristotle's Poetics*, Englewood Cliffs, N. J.; *Prentice-Hall, Inc.*, 1968, 112–136 in particular; Francis Ferguson, *Aristotle's Poetics*, New York, 1961.

Golding, John, "Two Who Made a Revolution: Picasso and Braque: Pioneering Cubism," *Museum of Modern Art*, New York, 1989.

Goldman-Rakic, Patricia, "Neurobiology of Mental Representation: The Mind, the Brain, and Complex Adaptive Systems," in *The Mind, The Brain, and Complex Adaptive Systems*, Eds., Morowitz, Harold J., and Singer, Harold L., Vol. XXI, Reading, MA: Addison-Wesley Publishing Co., 1995.

Gould, Jay, *Wonderful Life: The Burgess Shale and the Nature of History* (New York: Norton, 1990); for commentary on Gould and Darwin, see R. C. Lewontin, "Fallen Angels," *New York Review of Books*, Vol. XXXVII. No. 10, June 14, 1990.

Hawking, Stephen, *A Brief History of Time*, New York: Bantam Books, 1988.

Hayles, Katherine N., *Chaos and Order: Complex Dynamics in Literature and Science*, Chicago: The University of Chicago Press, 1991.

Heidegger, Martin, cited in Stephen Erickson, *Language and Being: An Analytical Phenomenology*, New Haven: Yale U. Press, 1970.

Hellerstein, David, "Plotting a Theory of the Brain," *The New York Times Magazine*, 23 May, 1988. Interview and review of Edelman, *Neural Darwinism*.

Holland, John, "Can There be A Unified Theory of Complex Adaptive Systems?" in *The Mind, The Brain, and Complex Adaptive Systems*, eds., Morowitz, Harold J., and Singer, Harold L., Vol. XXI, Reading, MA: Addison-Wesley Publishing Co., 1995.

Honzl, Jindrich, "Dynamics of Sign in the Theatre," trans. Susan Larson, in *Semiotics of Art: Prague School Contributions*, ed. Ladislav Matejka and Irwin Titunik, Cambridge, Mass., 1976.

Horgan, John, "Quantum Philosophy," *Scientific American*, July 1992.

Hsiang-Kuang, Chou, *The History of Chinese Culture*, Allahabad: Central Book Deposit, 1960.,

Isaacson, Robert L., *The Limbic System*, New York: Plenum Press, 1974.

Jarry, Alfred, *Gestes et Opinions Du Docteur Faustroll, Pataphysician*, "roman néo-scientifique," suivi de "Speculations, Pantagruel (with Eugène Desmolder,) opéra bouffe en cinques actress," with music by Claude Terrasse. Cited by Simon Watson Taylor, *The Ubu Plays*, New York: Grove Press, Inc., 1968.

Jastrow, Robert, *The Enchanted Loom*, New York: Simon and Schuster, 1981.

Jefferson, Ann, "Structuralism and Post-Structuralism," in *Modern Literary Theory*, eds., Ann Jefferson and David Robey, Totowa, N.J., 1982.

Jerison, Harry J., *Evolution of the Brain and Intelligence*, New York: Academic Press, 1973.

Johanson, Donald and Edgar, Blake, *From Lucy to Language*, London: Weidenfeld and Nicolson, 1996.

Johansen, J. Dines, et all, *Approaching Theatre*, Bloomington: Indiana U. Press, 1991.

Jones, Edwin, *Reading the Book of Nature: A Phenomenological Study of the Creative Expression in Science and Painting*, Athens: Ohio U. Press, 1989.

Kaplan, Stuart R., *The Encyclopedia of Tarot*, Stamford, CT., U. S. Games Systems, Inc. 1971.

Kauffman, Stuart, *At Home in the Universe: The Search for Laws of Self-Organization and Complexity*, New York: Oxford University Press, 1995.

Keith. A. B., *The Sanskrit Drama in its Origin, Development Theory and Practise* Oxford: Oxford U. Press, 1924.

Kelso, J. A. Scott, *Dynamic Patterns*, MIT Press, 1995.

Kinney, Hannah, et all, "Neuropathological Findings in the Brain of Karen Ann Quinlan: The Role of the Thalamus in the Persistent Vegetative State," *The New England Journal of Medicine*, Vol 20, No 21 (May 26, 1994), 1469–75.

Kirby, Michael, "Nonsemiotic Performance," *Modern Drama* 21 (March 1982).

Knapp, Bettina, "Antonin Artaud and the Theatre of Cruelty," in Docherty, Brian, Ed., *Twentieth-Century European Drama*, New York: St. Martin's Press, 1994.

Knapp, Bettina, *Antonin Artaud: Man of Vision*, New York: Avon Books, 1969.

Knowlson, James, catalogue: *Samuel Beckett: An Exhibition*: Reading University Library, May-July 1971, London: Turret Books, 1971.

Kola, Rocky, *Blind Watchers of the Sky* (New York: Oxford U. Press).

Kowzan, Tadeusz, "The Sign in the Theatre," *Diogenes* 61 (1968).

Kramer, Hilton, "Beyond the Avant-garde," *The New York Times, Magazine*, November 4, 1979.

Kuhn, Thomas S., *The Structure of Scientific Revolutions*, 2nd edition (Chicago: University of Chicago Press, 1970).

Langer, Susan, *Feeling and Form*, (New York, 1953).

Lewontin, R. C., "The Dream of the Human Genome," *The New York Review of Books*, Vol XXXIX, No. 10 (May 28, 1997.

Lieberman, Philip, *Uniquely Human: The Evolution of Speech, Thought, and Selfless Behavior* Harvard: Harvard University Press, 1991.

Livio, Mario, *The Accelerating Universe* (New York: John Wiley).

Lloyd, Seth, "Learning How to Control Complex Systems," *Bulletin*, The Bulletin of the Santa Fe Institute, Spring 1995, Vol. 10 No 1.

Lorenz, Malkus, Spiegel, Farmer, et al, "Deterministic Nonperiodic Flow," *Journal of the Atmospheric Sciences*, 20 (1963).

Lyons, Charles, "Character and Theatrical Space," in *Theatrical Space: Themes in Drama 9*, ed. Redmon, James. London: Cambridge U. Press, 1987.

MacLean, Paul, "The Triune Brain, Emotion and Scientific Basis," in F. O. Schmidt ed., *The Neurosciences, Second Study Program* (New York, 1970).

Madison, G. B., *The Hermeneutics of Postmodernity Figures and Themes* Bloomington: Indiana U. Press, 1988.

Marinetti, Filippo, "Beyond Communism," 1920, in *Marinetti: Selected Writings*, ed., R. W. Flint, Trans., R. W. Flint and Arthur Coppotelli, New York: Farrar, Strouss and Giroux, 1972.

———, "Le Futurism," *Le Figaro*, 20 February, 1909,. p. 1, "Futurist letter circulated among cosmopolitan women friends who give tango-teas and Parsifalize themselves," 11 January, 1914, "Dynamic and Synoptic Declamations," 11 March 1916, Emilio Settinelli and Bruno Corra, "The Futurist Synthetic Theater," 11 January 1915, 18 February 1915, "Manifesto of the Futurist Dance," 8 July 1917, "The Futurist Symphony," 11 September 1916, in *Marinetti: Selected Writings*, ed., R. W. Flint, Trans., R. W. Flint and Arthur Coppotelli, New York: Farrar, Strouss and Giroux, 1972.

Martin, Marianne, *Futurist Art and Theory 1909–1915*, Oxford: Clarendon Press, 1968.

Martin, Robert D., *James Arthur Lecture, Development of the Human Brain*, 1982., Cambridge: Harvard U. Press, 1973.

McAuliffe, Kathleen, "Get Smart: Controlling Chaos," *Omni*, Vol. 12, No. 5. February 1990.

McConachie, Bruce, "Towards a Postpositivist Theatre History," *Theatre Journal*, 37 (1985).

McDermot, Douglas, "The People's Institute of New York City," *Theatre Survey*, May 1965.

Mielin, Donald, *Origins of the Modern Mind: Three Stages in the Evolution of Culture and Cognition*. Cambridge: Harvard University Press, 1991.

Mishkin, Mortimer, and Appenzeller, Tim, "The Anatomy of Memory," *Scientific American* Off print, 1987.

Mounin, Georges, Introduction a la semiologie (Paris, 1970), 87, paraphrased in Carlson, Marvin, *Theories of the Theatre: A Historical and Critical Survey, from the Greeks to the Present*, Expanded Edition, 1993. New York: Cornell U. Press, 1993.

Mukarovsky, Jan "Art as Semiotic Fact," in *Semiotics of Art:: Prague School Contributions*, eds., Ladislav Matejka and Irwin R. Titunik, Cambridge, MA.: The MIT Press, 1976.

Mukarovsky, Jan, "On the Current State of the Theory of the Theatre," trans. Paul Garvin, in *A Prague School Reader on Esthetics, Literary Structure and Style*, Washington, 1955.

Mukarovsky, Jan, *Structure, Sign and Function*, trans. John Burbank and Peter Steiner (New Haven, 1978).

Nietzsche, Friedrich, *Selected Letters of Friedrich Nietzsche*, ed., and trans. Christopher Middleton, Chicago: University of Chicago Press, 1969.

Nietzsche, Friedrich, *Thus Spoke Zarathustra*, trans. Walter Kauffmann, New York: The Viking Press, 1966.

Ogden, Dunbar, "The Mimetic Impulse or the Doppelgänger Effect," ed. by Robert Erenstein, *Theater and Television*, Amsterdam: International Theatre Bookshop, 1988.

Old, Walter Gorn, *The Shu King, or the Chinese Historical Classic*, London: Theosophical Publishing Society, 1904.

Park, David, *The Fire Within the Eye* (Princeton: Princeton U. Press).

Pavis, Patrice, *Languages of the Stage*, trans. Susan Melrose, New York, 1982.

Peirce, C. S., *Collected Papers*, Cambridge, Mass., 1931–58.

Rees, Martin, *Just Six Numbers* (New York: Basic Books).

Ricoeur, Paul, *Husserl, An Analysis of His Phenomenology*, trans. Edward G. Ballard and Lester E. Embrec, Evanstown, Ill.: Northwestern U. Press, 1967.

Roach, Joseph R., *The Player's Passion: Studies in the Science of Acting*, Newark: University of Delaware Press, 1985.

Roach, Joseph, "Problems and Prospects for Theatre Research," Statement to the American Council for Learned Societies, undated, in response to a charge from Milton Gatch, to ACLS Delegates on September 11, 1992.

Rosenfield, Israel, "Neural Darwinism: A New Approach to Memory and Perception," *The New York Review of Books*, Vol XXXIII, No. 15 (October 9, 1986).

Ross, Phillip, "Hard Words," *Scientific American*, Vol. 264, No. 4 (April, 1991).

Rouse, John, in Janelle Reinelt, "Semiotics and Deconstruction: an Introduction," *Critical Theory and Performance*, eds. Janelle G. Reinelt and Joseph Roach, Ann Arbor: The University of Michigan Press, 1992.

Rubin, William, "Picasso and Braque: An Introduction," *Picasso and Braque: Pioneering Cubism*, New York: The Museum of Modern Art, 1989.

Russell, John, "Modernism to Postmodernism: A New World Once Again," *The New York Times*, Section 5, August 21, 1982.

Schechner, Richard, *Performance Theory*, Revised and Expanded Edition, New York: Routledge, 1988.

Schrodinger, Erwin, *What Is Life? The Physical Aspect of the Living Cell*, Cambridge, U.K.: Cambridge U. Press, 1955.

Schumacher, Artaud, xxx, Citation from Marowitz, Charles, *Artaud at Rodez*, London: Marion Boyars, 1977.

———, ed. and trans., and Singleton, Brian, trans, *Artaud on Theatre*, London: Methuen, 1989.

Scott, A. C., *The Classical Theater of China*, Westport: Greenwood Press, 1957.

Shakhar, Indu, *Sanskrit Drama: Its Origin and Decline*, Leiden: E. J. Brill, 1960.

Shattuck, Roger, *The Banquet Years: The Arts in France 1885–1918*, New York: Harcourt Brace, 1958.

Shaw, George Bernard, "Some Particulars by Shaw," *The Quarterly Review*, No. 3 (Winter 1946).

Sherrington, C. S., *Man and His Nature*, Cambridge, U.K. Cambridge U. Press, 1951.

Simon, Neil, and David Rabe, "The Craft of the Playwright," ed. Samuel G. Freedman and Michael Williams, *The New York Times Magazine*, May 26, 1985.

Smith, John Maynard, "Taking a Chance on Evolution," *New York Review of Books*, Vol. XXXIX, No. 9, May 14, 1992.

Sokel, Walter H., *Anthology of German Expressionist Drama*, Garden City, N.J., Doubleday and Company, Inc., 1963.

Sontag, Susan, *Antonin Artaud: Selected Writings*, ed., and intro., Susan Sontag, New York: Farrar, Strauss and Giraux, 1976.

Spencer, Charles E., *The World of Serge Diaghilev*, New York: Penguin Books, 1979.

States, Bert O., *Great Reckonings in Little Rooms*, Berkeley, U. of California Press, 1985.

———, "The Actor's Presence: Three Phenomenal Modes," *Theater Journal*, 35, 1983.

Stone, Lawrence, "The Revolution Over the Revolution," *New York Review of Books*, Vol.XXXIX, No. 11, June 11, 1992.

Tisdale, Caroline, and Angelo Bozzola, *Futurism*, New York: Oxford University Press, 1978.

Turner, Victor, "Body, Brain, and Culture," *The Anthropology of Performance*, Performing Arts Journal, 1986.

Turner, Victor, *Drama, Fields and Metaphors: Symbolic Actions in Human Society, Myth and Ritual* Series. Ithaca: Cornell University Press, 1974, and *From Ritual to Theatre: The Human Seriousness of Play* (New York: Performing Arts Journal, 1982). There are significant other Turner publications, acknowledge by Schechner in Performance Theory: "Body, Brain, and Culture," *Zygon 18* (3): 221–45; *On the Edge of the Bush*, Tucson: University of Arizona Press, 1985; and *The Anthropology of Performance*, New York: *Performing Arts Journal*, 1986).

Turner, Victor, "Frame, Flow, and Reflection: Ritual and Drama as Public Liminality," *Performance in Postmodern Culture*, Madison: Coda Press, 1977.

Turner, Victor, *From Ritual to Theatre*, New York, 1982.

Vajda, György, "Outline of the Philosophical Backgrounds of Expressionism," *Expressionism as an International Literary Phenomenon*, ed., Ulrich Weisstein (Paris: Librarie Marcel Didier, 1973).

Valéry, Paul, *Collected Works, Vol. VII*, ed. Jackson Mathews, Trans. Malcolm Cowley and James Lawler, (New York: Princeton U. Press, n.d.).

Varadpanda, M. L., *Traditions of Indian Theater*, New Delhi: Abhinar Publications, 1978.

Vince, Ronald W., "Issues in Theater Historiography," paper presented at the XIth World Congress of the International Federation for Theater Research, Stockholms universitet, Stockholm, 31 May, 1989.

Weinberg, Steven, *The First Three Minutes* (New York: Basic Books).

Williams, David, Ed., *Peter Brook and the Mahabharata: Critical Perspectives*, New York: Routledge, 1991.

Wilshire, *Role Playing and Identity: The Limits of Theater as Metaphor*, Bloomington: Indiana U. Press, 1982.

Wilson, E. O., *Sociobiology: The New Synthesis*, Cambridge: Harvard University Press, 1975.

Wilson, Robert, "Theaterproben IX Rehearsals," *The Forest Playbook*, n.p., n.d.

Wilson, Wilson, unpublished application for funding of *the CIVIL warS*, for the 1984 Los Angeles Olympic Arts Festival, American Repertory Theatre archives, Cambridge, Massachusetts.

Wimsatt, W. K. Jr., and M. C. Beardsley, "The Intentional Fallacy," *Sewanee Review, 54*, 1946.

Wimsatt, W. K. Jr., "Genesis: A Fallacy Revisited," *The Disciplines of Criticism*, ed. Peter Demetz, New Haven: Yale U. Press, 1968.

Index

neural theatricalism, 30
    definition of, 38
neuroanatomy, 86
New Critics, 36
New Historicism/Theatrism, 48
New Theatrism, 47
New York Times, 103
Newton, I., 44
Nichols, J., 69
Nietzsche, F., 42, 49, 51, 53, 84, 98, 99
Norman, M., 32, 105
Not I (Beckett), 124

Old, W. G., 134
Opposing Mirrors and Video Monitors on Time Delay, 118
ordered complexity, 62
Origin of Species (Darwin), 65, 97
Ortolani, B., 106
Other, 93, 101

Pabo, S., 8
Paradoxe sur le Comèdien (Diderot), 32
Pavel, P., 15
Pavis, P., 52
Pear Orchard Conservatory, 137
Peirce, C. S., 52
Penrose, R., 114
Performance Studies, 47, 75
perspectivism, 36
Picasso, P., 38, 61, 72, 85
Pirandello, L., 87, 105
Plato, 113
Play (Beckett), 119, 122
Prague Linguistic Circle, 35
Prague Structuralists, 78
Player's Passion, The (Roach), 32
Poincare, H., 44, 112
positivism, 104
Prigogine, Ilya, 45
Primary Demonstration: Horizontal-Vertical, 118

Proust (Beckett), 14
Proust, M., 71, 128
Ptolemaic asymmetry, 32
punctuated equilibrium, 11, 74, 121
Puranas, *132*

quadrune brain, 84
    theatrical brain, 88–89
quantum biology, 47
Quantum Mechanics, 18, 44
Quantum Theory, 41
Quinlan, K. A., 59, 61

Rabe, D., 101, 103, 104
Racine, Jean, 110
radical subjectivity, 36, 105
Ramayana, 132
Rasa, 132
Recer, P., 23
Rieder, R. D., 56
Roach, J., 32, 117, 119
Romeo and Juliet, 93
Ross, P., 21
Rouse, 117
Royal Shakespeare Company, 51
Russian formalism, 52

San Francisco Mime Troupe, 93
Sankai Juku, 107
Sanskrit drama, 132
Santa Fe Institute, 4
Saussure, F. de, 52
Schechner, R., 33, 35
Schrodinger, E., 18, 118, 119, 121, 122, 123, 125, 126, 127, 128, 129, 139
Schrodinger, I., 47
Schumacher, C., 50, 56, 58
Science Journal, 109
Scientific American, 14
Scott, A. C., 134
Scott, V., 43